ARKANSAS WILDLIFE

ARKANSAS WILDLIFE

A History

Arkansas Game and Fish Commission
Steve N. Wilson, Director

Edited by Keith Sutton

THE UNIVERSITY OF ARKANSAS PRESS
FAYETTEVILLE 1998

Manufactured in the United States of America

02 01 00 99 98 5 4 3 2 1

Designed by Liz Lester

♾ The paper used in this publication meets the minimum requirements of the American National Standard for Permanence of Paper for Printed Library Materials Z39.48-1984.

LIBRARY OF CONGRESS CATALOGING-IN-PUBLICATION DATA

Arkansas wildlife : a history / Arkansas Game and Fish Commission; edited by Keith Sutton.

p. cm.

Includes bibliographical references and index.

ISBN 1-55728-536-5 (hardcover : alk. paper). — ISBN 1-55728-537-3 (pbk. : alk. paper)

1. Game and game-birds—Arkansas—History. 2. Zoology—Arkansas. 3. Hunting—Arkansas—History. 4. Fishing—Arkansas—History. 5. Wildlife management—Arkansas—History. 6. Arkansas—History. I. Sutton, Keith. II. Arkansas Game and Fish Commission.

SK53.A754 1998

333.95'4'09767—DC21 98-26949

CIP

r98

Contents

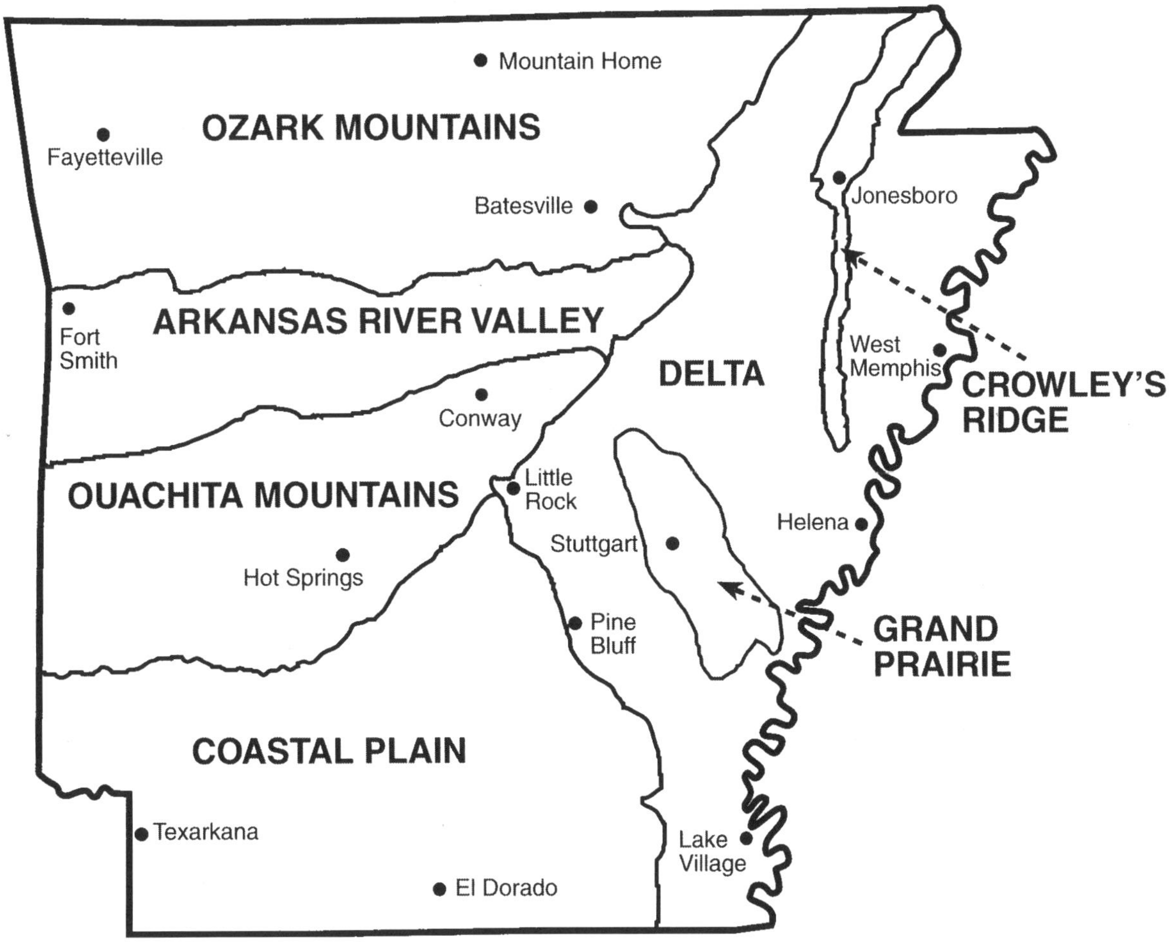

PART I

Discovery and Settlement

I Early Exploration

March 31—

[W]e set out at six o'clock in the morning; at ten o'clock, our Savage killed four turkeys and a buffalo for us, which forced us to stop to smoke them this day . . . this afternoon, our Savage killed a buffalo and a turkey for us . . .

April 1—

[L]ast night there was much rain. We left camp at six o'clock in the morning. Having advanced three-quarters of a league, we landed to change ourselves. During this time, two buffaloes crossed the river; we ran ahead over the shoal which made them turn in their tracks away from the other side . . . We killed them at their coming out of the water; after having quartered them, we continued our voyage.

April 11—

[L]ast night we heard a great number of wolves around our camp . . .

[W]e camped on the right, near a shoal, upon which Bontemps, Mr. Bresent, and I killed two buffalo bulls and four cows and took a calf alive. It is a remarkable thing that these wild buffaloes go in herds, and that all of each herd compose one family, so that, when they have not scented the hunter and when he is able to kill the old bull and the old mother of the herd, none of the herd takes flight, being accustomed to following their elders as being the most experienced. Thus, in this way, they all commit themselves to the killing one after another . . .

We are commencing to find stags and hinds; turkeys are common, in like manner are deer and bear; these lands are always very beautiful, forming small hills and valleys, diversified by plains and small mountains.

The river continues to widen; . . . it is abounding with fish and the current is almost like that of the Mississippi . . .

April 14—

We killed three buffaloes.

April 15—

The lands continue to be beautiful, and the river very navigable . . . the current is moderate and game abundant.

> We killed two buffaloes and three turkeys today.
>
> *April 22—*
>
> [W]e remained in camp to smoke meat; we killed a bear and a cub bear, which we made come out of a hollow tree with fire.
>
> [A] prodigious number of water turtles on logs.

The year is 1721. These are excerpts from the journal (Smith, *Arkansas Historical Journal,* winter 1951) of Frenchman Bernard de la Harpe, and they speak eloquently to the beauty of Arkansas and to the abundance of life-sustaining wildlife that greeted the first white men who explored what was then a massive wilderness.

According to *Arkansas and Its People,* edited by David Y. Thomas, La Harpe, however, was a latecomer. Arkansas history began 160 years earlier with the coming of Hernando de Soto. The Spanish nobleman's quest for gold and silver brought him and his men to northern Mississippi by the winter of 1540–41. In spring, they marched northwest and came to the Mississippi River in May, perhaps near what is now the city of Memphis. Soon after, they became the first Europeans to set foot on Arkansas soil. It is impossible to trace de Soto's wanderings accurately through Arkansas. The expedition's records are meager. Rivers, mountains, and other landmarks were unnamed and left that way. Dates, distances, and directions were rarely given or were guesses. The names of Indian villages often were given, but native peoples, because of wars and migrations, did not have permanent settlements. Based on clues found in the meager records, historians believe de Soto's wanderings took him into southwest Arkansas as far as the Red River.

It appears from the expedition's journals, however, that de Soto arrived at an Indian village between the St. Francis and Mississippi Rivers in June 1541. One journal states that where de Soto stayed in the village was "a great lake, that came near unto the wall; and it entered into a ditch that went round about the town . . . From the lake to the great river was made a wear [weir] by which the fish came into it, which the cacique [chief] kept for his recreation and sport." The writer describes several kinds of fish. "There was another fish called the peel fish; it had a snout of a cubit long . . . and in the river there were some 150 pounds in weight, and many of them were taken with the hook" (Thomas 1930).

By European standards, the expedition was a failure because no precious metals were found. The Spaniards learned something of the country's geography—its rivers, mountains, soils, climate, and inhabitants. But they made no efforts to claim or settle the land.

A small band of Frenchmen were the next Europeans to come to Arkansas—131 years after de Soto. By 1665, the French had settled on the St. Lawrence River and rapidly were pushing their explorations and claims to the west and southwest. In spring 1673, Frontenac, one of the governors of New France, gave permission and the necessary help for an expedition to find the "Missipi" River. The expedition was commanded by a Jesuit priest, Father Jacques Marquette, and a trader, Louis Joliet. These men, accompanied by five rowers and two Indian guides (who later turned back), began their long journey in two canoes on May 13, 1673. In his journal, Marquette noted seeing beautiful flowers, vast plains, and shaggy buffaloes,

among other things. At one point, he said, "[W]e met from time to time monstrous fish, which struck so violently against our canoes that we took them to be large trees, which threatened to upset us" (Thomas 1930).

Even after the Marquette-Joliet expedition, which ended when it reached the mouth of the Arkansas River, nothing was done to occupy the Mississippi Valley. This was left to the next explorer to reach Arkansas—Sieur de La Salle. Frontenac and La Salle were bold and had ambitions to establish the fur trade. They planned to have a chain of forts built down the Mississippi and to plant colonies in the river valley. This would hold the country for France, control the Indians, and establish a monopoly in the fur trade.

La Salle left the Illinois country on his journey south in spring 1682. He was accompanied by, among others, Henri de Tonti, a French soldier who is often called the "Father of Arkansas." La Salle needed a soldier as lieutenant because he didn't propose to stop at the Arkansas as Marquette and Joliet had done but to go to the very mouth of the Mississippi River, among the Spaniards if necessary, and claim the entire country for France.

The La Salle expedition stopped at de Soto's old crossing at Chickasaw Bluffs near Memphis. Then they went to an Indian village at the mouth of the Arkansas River. Here La Salle took formal possession of the territory for his king, Louis XIV, and in his honor called it Louisiana.

De Soto discovering the Mississippi River

In May 1541, Hernando de Soto's quest for gold and silver brought his expedition to the Mississippi River, perhaps near what is now Memphis. It appears from the journals kept by his men that de Soto's party arrived at an Indian village between the St. Francis and Mississippi Rivers in June 1541, thus becoming the first Europeans to set foot on what was to become Arkansas.

Arkansas History Commission.

Sieur de La Salle

La Salle taking possession of land at the mouth of the Arkansas River, March 13, 1682.

Arkansas History Commission.

As he had intended, La Salle traveled on to the mouth of the Mississippi, then returned to Quebec. Subsequently, he went to France seeking additional government support to establish a colony near the mouth of the mighty river. On his return to Louisiana, La Salle went off course and was murdered from ambush.

De Tonti, his lieutenant, was leading a party down the Mississippi to meet the returning La Salle. After a long fruitless search for the explorer, de Tonti went back up the Mississippi River and stopped en route to visit his old friends, the Arkansas Indians (the Quapaw). La Salle earlier had granted some land near the mouth of the Arkansas River to de Tonti. Some of de Tonti's party, pleased with the country, the friendliness of the Indians, and its good location for a trading post, begged him to let them stay. In 1686, de Tonti gave permission for ten of his men to establish a fort, trade with the Indians, cultivate the land and make an effort to civilize the natives. He built his own store at an Indian village called Assotone, a short distance from the mouth of the Arkansas. Arkansas Post was the first settlement of its kind in the middle or lower Mississippi Valley (Thomas 1930). Early settlers in the Mississippi Valley did poorly. Malaria and yellow fever killed many. Nothing was done to settle the interior of the country, and de Tonti's post on the Arkansas was neglected. However, the founding of Arkansas Post opened the door to commercial hunting in the middle reaches of the Mississippi River Valley. French hunters and trappers, employing Indians, ranged far up the Arkansas, White, and St. Francis Rivers to kill deer and furbearers.

The first real attention Arkansas received was in 1712 when Louis XIV, having become exasperated with the expense of Louisiana, turned control of the vast terri-

Henri de Tonti, a French soldier, is often referred to as the Father of Arkansas. In 1668, de Tonti gave permission for ten of his men to establish a fort near the mouth of the Arkansas River, trade with the Indians, cultivate the land, and make an effort to civilize the natives. De Tonti built a store of his own at an Indian village called Assotone, a short distance from the mouth of the Arkansas River. This trading post, later known as Arkansas Post, was the first settlement of its kind in the middle or the lower Mississippi River Valley.

Arkansas History Commission.

Lithograph of early dwellings and boat

tory over to a private company. But the company's owner quickly became convinced the area was worthless and gave up the charter. It was picked up by John Law, a Scotsman who was in France at the time. He organized the Western Company with the intent of colonizing the land, even if it took deceptive means to do it. Law obtained a twelve-mile-square tract of land on the west side of the Arkansas River above Arkansas Post. For this grant, Law was required to settle fifteen hundred persons on the land and to keep soldiers there to protect the colony.

Henri de Tonti

Law went bankrupt and left France in the latter part of 1720. Storehouses and homes in and around Arkansas Post soon were deserted. The Western Company did not want to lose all Law accomplished, so it appointed a director for Arkansas Post. One of his duties was to obtain colonists to settle there. A few soldiers from New Orleans also were stationed there for many years, and the settlement continued as a trading post. Boatloads of supplies and products passed yearly between the Illinois country and lower Louisiana. Robber bands, composed of whites and Indians, preyed on this trade near the mouth of the Arkansas River, so a small fort and storehouse for supplies were built.

The Western Company sent several expeditions to explore the area. One was La Harpe's. He was ordered to take sixteen soldiers, to start from Arkansas Post and to ascend to the headwaters of the Arkansas River. He was to observe the quality of the land and learn about the Indian tribes and mines (if any), keep a journal, note the navigability of the stream, and ascertain whether the Spaniards had any settlements in the region. La Harpe left Arkansas Post on March 10 and proceeded up

Early French trading post

Arkansas Post on the lower Arkansas River, 1686.

Arkansas History Commission.

the Arkansas River until April 17. Sickness among his men compelled La Harpe to abandon the expedition at that point and return to Arkansas Post. While he failed to get all the information his governor wanted, La Harpe documented in his journal that Arkansas was a beautiful land with bountiful wild game.

Jean Bernard Bossu, captain in the French Colonial Marine who visited Arkansas Post in 1751 and again in 1770–71, described a buffalo hunt the Quapaw Indians made some distance from the post, perhaps on the Grand Prairie. The Indian men rode horses and carried long lances tipped with crescent-shaped iron blades. A rider would overtake a buffalo, hamstring it with a lance, and dash on to the next animal while women and boys clubbed the crippled buffalo to death. French records of the 1700s speak of buffaloes along the Boeuf River, which was named for the bison, in what is now Chicot County; on the Saline River in the Ouachita Valley; occasionally on the St. Francis River; and on the Arkansas River. The last buffalo herd in south Arkansas was killed in the Saline River bottoms in 1808. A group of Frenchmen from Louisiana joined local settlers and slaughtered 150 buffaloes. They scoured the woods to get every last one and then, after removing the choice cuts for a celebration, left the carcasses to the vultures (Holder 1951).

In 1796, the young United States government began establishing what was called a factory system to control trade with the Indians. Government-owned and -operated trading houses were opened in frontier areas for use by Indians, with goods and supplies being bartered for furs and skins. The stated purpose, according to historian Aloysius Plaisance, was to "gain the friendship of the Indians by supplying them with necessities at a low price and to overcome the influence of the British and Spanish trappers."

In "The Arkansas Factory, 1805–1810," Plaisance details the business of such trading houses, including one established at Arkansas Post after the United States made the Louisiana Purchase in 1803. The private American Fur Company led

Early French and Spanish explorers encountered many groups of native Americans when they arrived in Arkansas. By the late 1700s, several private companies had established trade with tribes like the Osage, bartering goods and supplies for furs and skins taken by the Indians.

Arkansas History Commission.

Indian community

Osage Indians with game in canoe

Spanish adventurers in Indian village

Bison

Bison, or buffaloes, once roamed throughout much of Arkansas. They were wiped out in the state by 1837, perhaps the earliest victims of overzealous hunters.

Arkansas Game and Fish Commission.

opposition to this government competition. It appears, however, that economics was a greater factor in the system's demise in 1822. "The overwhelming number of deer skins that glutted the skin market" caused the system to lose money.

The existence of the factory system did not keep the United States War Department from granting permission to private houses or firms to do business with the Indians. The House of Bright and Morgan, for example, received permission to trade with the Osage Indians. The factory system also did not impinge on the success of the private companies. In spring 1806, 975 packs of deer skins were shipped out of Arkansas Post. Of this total, 267 packs belonged to the House of Bright and Morgan, whereas only 61 were from the factory. Bright and Morgan also had 1,500 pounds of beaver skins, 930 bear skins, and a "considerable quantity" of bear oil and tallow.

According to Plaisance, John B. Treat operated the Arkansas Post factory. His books reveal that in spring 1806, prices for various furs and peltries were $.40 a pound (French weight) for shaved deer skins, $.30 a pound for deer skins in hair, and $1.00, $1.50 and $2.00 each for bear skins. Treat invoiced all bear skins he received at $1.50 each and paid $1.25 a pound for beaver and $.25 each for raccoon, bobcat, and fox pelts. In his first year's report, Treat commented that his collection of furs was far greater than he anticipated, particularly considering the opposition of local trappers to the factory system: "This quantity . . . makes upward of 170 Packs which has been received at the Factory, a collection far greater than the most sanguine expectations led us to suppose would be procured."

Treat's handling of deer skins the Indians brought him, however, showed his inexperience in dealing with furs and skins. They were bundled together and shipped by boat to New Orleans in care of a government agent there. The agent, Joseph Saul, told Treat by letter that many skins were damaged or rotten, and he gave the manager directions for packing and shipping them correctly. The Arkansas Post factory reached its peak of activity in 1807, after which it went into a steady decline until it closed in 1810.

Southwest Arkansas also became the scene of intense hunting and trapping activities during the 1700s and early 1800s. The Ouachita, Little Missouri, and Red Rivers and other smaller streams were hunted by Frenchmen from the Post of Ouachita, where Monroe, Louisiana, now stands. From the Post of Natchitoches, Louisiana, came Champagnolle, Tulipe, and several other hunters who gave their names to places in the Ouachita River area (Holder 1951). When William Dunbar and George Hunter explored the Ouachita River region in 1804, they found that hunters went three hundred miles upriver in search of game, and there was evidence Frenchmen had lived along the nearby Red River for fifty years.

Great quantities of skins and bear oil sewed up in deer skins came down the streams, finding a ready sale in New Orleans, where bear skins sold for one dollar and oil for one dollar a gallon. An early settler south of Columbia County remembered seeing large parties of French Creoles from around Natchitoches going north for their annual bear hunt as late as the 1820s. They went in December, returning usually in February, their ponies laden with skins and meat (Martel 1943).

Homesteaders and wagon train

President Thomas Jefferson's acquisition of the Louisiana Purchase brought a flood of eastern homesteaders who decided to make new beginnings "out west." Settlers soon arrived in large numbers in Arkansas, which was then part of the Missouri Territory.

Arkansas History Commission.

2 Settling the Wilderness

President Thomas Jefferson's acquisition of the vast lands known as the Louisiana Purchase was like removing a finger from a hole in a dam. Many in the eastern United States were beginning to feel their area was becoming too crowded, or they had worn out their land through poor farming practices and needed to search for homesteads to make new beginnings. Settlers began arriving in greater numbers in Arkansas, which was then part of the Missouri Territory.

John Billingley's father brought his family to Arkansas from middle Tennessee in 1814. Their first stop was Arkansas Post. The Billingsley family traveled up the Arkansas River to its junction with Cadron Creek. After a year at Cadron, the family moved farther west—to Big Mulberry. In 1816, Mulberry consisted of about thirty families. Billingsley recalled that he and his family lived there two years "enjoying all the luxuries of life that a new country could afford, such as buffalo, bear, deer, elk and fish and honey." The family's substitute for bread was "venison dried by the fire and then pounded in the mortar and made up in small cakes and fried in bear's oil." The French came up the Arkansas River in large canoes, Billingsley said, "and supplied us with domestic [cotton cloth] and checks [checked cloth] and earthing ware and calico . . . This was all paid in bear skins and deer skins and coon skins and bear oil, some beavers and otter skins and bees wax and that in abundance. For we had honey in any amount" (Worley, "Letters from an Early Arkansas Settler" 1952).

John C. Benedict with three of his brothers-in-law came to central Arkansas in 1811. After a stop in Little Rock, they set up camp south of the mouth of the Devil's Fork of Little Red River where they cleared about thirty acres of land and built three

Early fur trappers with coonskins, including lady in coonskin clothes

Prior to 1823, trappers used deadfalls and snares to capture furbearers such as raccoons. The invention of the leg-hold trap that year greatly increased trappers' effectiveness. Animals thus caught provided meat for the table, furs for clothing, and a medium of exchange when pelts were traded for goods.

Arkansas Game and Fish Commission.

small cabins. The group remained at this settlement for about two and a half years, subsisting almost entirely on the flesh of buffaloes, bears, elk, deer, and turkeys.

Also in the vicinity was the Flanagin family, made up of the families of two very old men, "Old Hugh" and "Old Bill," who had five or six sons each, making about fourteen men, and several married daughters. The sons and daughters had large families of their own. "This family . . . knew nothing about, nor did they pretend to farm. They knew or cared for nothing but their Guns and Dogs and passed their days in hunting and trapping for Beavers, Otters, killing Buffalo, Bear, Elk, and Deer, and like the Indians studiously endeavoring by every means in their power to discourage immigration and to injure newcomers" (Worley 1951).

Henry Rowe Schoolcraft and a companion, both inexperienced hunters, provided additional testimony about the abundant game greeting Arkansas's early settlers. They subsisted "entirely on wildlife during several months of tramping over southern Missouri and northern Arkansas." During two months traveling the Arkansas Ozarks, Schoolcraft recorded twenty species of game, ranging from quail to buffalo. He related that "on Christmas morning, 1818, a hunter of the upper White River killed 14 turkeys in two hours" (Worley, "An Early Arkansas Sportsman" 1952).

According to Glenn G. Martel, a Columbia County historian, many settlers on their way to Texas stopped in southwest Arkansas because of inclement weather, sickness, weariness of travel, the sight of a beautiful wooded home site near a spring, a good range for their cattle, or an abundance of game.

Hunters not only furnished meat for the table, but peltries were the medium of exchange, taking the place of cash. Furbearers were abundant, but trapping them was hard work. The earliest white trappers used Indian techniques, primarily snares and deadfalls requiring tremendous amounts of time and labor to build. They were inefficient, often failing to hold the animal. It wasn't until 1823 that Sewell Newhouse revolutionized trapping with the leg-hold trap, which added greatly to trappers' effectiveness because it could be used for long-legged animals such as wolves and foxes, as well as for beavers, muskrats, minks, otters, and other short-legged animals.

In addition to food and "cash" in the form of peltries, settlers frequently relied on wildlife for clothing. From interviews conducted on November 11, 1932, with

John Murphy and on October 28, 1932, with W. V. Keith, two early Columbia County settlers, Glenn Martel learned how Arkansas pioneers obtained the shoes and boots they wore. A Charles Hodge was a shoemaker who tanned leather at the Rogers water mill on Smackover Creek. Murphy said Hodge "came often to our house and took the measure of our feet; then he went back to his shop and made the shoes. It would usually be a long time after we made the order before we got the shoes. Gloves and sometimes shoes were made of fox squirrel skins; they were superior to kid in durability and equal in quality." Keith told Martel that John Key, another local tanner, made him a pair of dress shoes. "He sent me into the woods with a rifle with instructions to kill a number of fox squirrels, taking care to shoot them in the head so as not to spoil the skins. By boiling them in 'ooze' [extract of bark], skins were tanned in a week. Three skins were necessary for each shoe. This was the best pair of shoes I ever had."

Southwest Arkansas, Martel said, was covered with a heavy growth of pine and hardwood trees and a dense undergrowth that made traveling through the forest difficult. Fires, however, cleared some of the undergrowth, leaving an open country "with a luxuriant growth of knee-high tender, nutritious grasses, wild pea vines, and switch cane—an ideal range for the deer and turkey and later for the settlers' livestock."

The presence of so many bears, wolves, panthers, and bobcats, however, made raising livestock a precarious venture for settlers. Or at least they thought it did. Settlers viewed red wolves as the "great predator on small stock," yet wolves also helped create a sense of community among pioneering families (Martel 1943). Near dark, wolves began howling, at first piteously and dolefully, making the most cheerful household lonely. There would be two or three wolves at different points at first, then others would join in until the pack seemingly would increase to forty or fifty. As they joined in, the howls increased until they became an "indescribable medley of whining, yelping, yelling, howling, discordant sounds that would make the hair rise on one's head." Then the pack would go silent for four or five minutes before repeating the performance. This would continue for about two or three hours. When wolves began howling, each family started blowing a horn. Some families had two or more horns. The horns caused the homestead dogs to howl, which the settlers thought would keep the wolves at bay. The horn blowing became "a kind of roll call and concert, and if any family failed to join, messengers would soon be at their doors to know if there was anything the matter" (Neal 1958).

Because settlers perceived the predators as a problem in raising livestock, the Missouri territorial legislature enacted a bounty law on December 23, 1816, designed to "encourage the killing of Wolves, Panthers, and Wild-Cats." Almost two years later—on December 11, 1819—the legislature repealed the law ("Records of the States").

Arkansas became a territory separate from Missouri in 1819. President James Monroe appointed Gen. James Miller of New Hampshire, a hero of the War of 1812, as governor of the Arkansas Territory. In a letter written in 1820, Governor Miller boasted, "Of animals in this country, both winged and quadruped, we have no want. There is almost every species of bird and fowl in great abundance; wild geese and swans, turkeys, quails, rabbits, raccoons, bear, wolf, catamount, wild cat, beaver, otter, deer, elk, and buffalo—a huntsman has full scope" (Williams 1983).

Hunters with mountain lions

Predators such as the red wolf and mountain lion were killed on sight because they were considered competitors for game and dangers to the settlers' livestock. One early Arkansan wrote, "There were predatory animals, particularly bear, panthers, and almost innumerable bobcats and catamounts, and packs of large timber wolves, all of which inflicted heavy losses on the settlers by destroying hogs, cattle and colts" (Simonson 1947).

Arkansas Game and Fish Commission.

The Dunbar and Hunter expedition, when ascending the Ouachita River in 1804, saw a flock of swans near present-day Rockport in Hot Spring County. The birds were at the water's edge beneath overhanging rocks. As the boats approached, the swans floated out, making "tremulous, melancholy accents." Dunbar wrote that the scene was a "truly sublime picture" (Rowland 1930). Swans graced streams and lakes throughout Arkansas, and their plumage was prized highly by Indians. The chiefs wore beautiful capes made of swan skin with the feathers intact. Because they were overhunted, however, the majestic birds were almost gone from Arkansas by the time of the Civil War (Holder 1951).

Governor Miller wrote his boastful letter about the same time Thomas Nuttall, a renowned naturalist, was concluding a two-year journey through Arkansas that took him into what is now eastern Oklahoma. From time to time, Nuttall wrote in his journal about animal life in regions he visited. "Red birds," "blue birds," wild geese, and turkeys were noted near Arkansas Post, which then consisted of twenty or so scattered houses and several stores. Around the Cadron settlement on the Arkansas River, Nuttall found an abundance of "wild cats" (skunks) both striped and spotted, as well as panthers, bears, and wolves. "The bison (improperly called buffalo) is also met with occasionally. Deer, too, abound here as well as in other parts of the territory," he wrote (Vaulx 1946).

Nuttall wrote that ticks were "troublesome," especially after he ascended the Poteau River and crossed the rough, hilly terrain from the Winding Stair Mountains to the Kiamichi River, a tributary of the Red River. He noted some species of wildlife that would not catch the eye of the ordinary, untrained observer. For example, "Near the banks of the Red River, about the confluence of the Kiamesha," he commonly saw the "conspicuously beautiful" scissor-tailed flycatcher he already had observed nesting in an elm near Fort Smith (Graustein 1967).

Virgin prairie in Arkansas County

Prairie grasses covered only about 1 million of Arkansas's 34 million acres in the early 1800s. The rest were covered with trees.

Arkansas History Commission.

3 Territorial Days

With its separation from Missouri in 1819, Arkansas acquired political boundaries. That year, Nuttall described the state as "one vast trackless wilderness of trees" (Stroud 1981). It was an accurate observation, for trees and forests covered all but about a million acres where prairie grasses dominated.

Arkansas was renowned regionally at this time as a hunting, trapping, and fishing paradise. This reputation began growing nationally after Charles Fenton Mercer "Fent" Noland moved to Batesville from Virginia in 1826. Noland was the creator of Pete Whetstone of Devil's Fork who regularly regaled readers of the New York *Spirit of the Times* with his yarns about hunting and fishing adventures in Arkansas. Some of Noland's Batesville area neighbors kept packs of bear dogs and deer dogs, and he frequently accompanied them on long hunts to such places as Oil Trough Bottom, War Eagle Creek, and Devil's Fork. It is Noland who said Oil Trough got its name from the storage of bear oil in troughs made from hollow logs.

Noland told *Spirit of the Times* readers on August 27, 1836, that he recently had spent a night with a hunter who the day before had killed five deer in twenty-five minutes. On four trips into the woods in the winter of 1838, Noland said he saw at least fifteen deer on each trip. Of smaller game, such as turkeys and quail, Noland told the newspaper's readers on November 22, 1845, they were "thicker than blackberries." Noland also described the wingshooting of ducks, prairie chickens, and quail. Noland considered geese the hardest to kill among the small game, and quail the most fun to shoot and the best to eat.

In the February 21, 1852, issue of the newspaper, Noland boasted through his

C. F. M. Noland

Charles Fenton Mercer "Fent" Noland of Batesville was one of Arkansas's first outdoor writers. Under the pen name Pete Whetstone, Noland regaled readers of the *New York Spirit of the Times* with yarns about his adventures hunting and fishing in Arkansas. His stories and those of other writers such as Thomas Bangs Thorpe, Tom Breese, and William Quesenberry gave Arkansas a national reputation as a matchless hunting and fishing ground.

Arkansas History Commission.

fictional alter-ego Pete Whetstone that he had killed fifty-four turkeys in the first few days of a single season. As late as 1853, turkeys were still numerous on Poke Bayou about four miles from Batesville.

Noland's early enthusiasm in Arkansas was bear hunting. His bear yarns and those of other writers, such as Thomas Bangs Thorpe, Tom Breese, and William Quesenberry, gave Arkansas a national reputation as a matchless hunting ground of wild and woolly thickets and canebrakes. It was no accident, therefore, Arkansas's unofficial but popular name was "The Bear State."

His writings and those of Quesenberry, a Fayetteville resident, also gave Arkansas fame as a fishing state. Noland fished for anything from "bull minners to ponderous cat." His favorites, however, were bass and jack salmon (walleyes). At Jacksonport or Figlo Lake, it was easy, Noland reported, to catch a horse load of "salmon" in an hour or two. They were the most common game fish in the White River and sometimes reached weights of six or seven pounds, he wrote. One of his favorite bass fishing streams was the Cache River, where, it has been said, fishermen could catch from eighty to one hundred of the fish in a half day (Worley, "An Early Arkansas Sportsman" 1952).

Strangely, few, if any, published quotes from early settlers speak to the presence of alligators in Arkansas, though much of the eastern and southern parts of the territory were swampy. They inhabited the territory, though, for in May 1828 the *Arkansas Gazette* reported an eleven-foot alligator was killed just across the Arkansas River from Little Rock and that "nine rifle balls were fired into his tough hide before he died" ("Chips and Shavings" 1952).

Noland wasn't the only settler with tales to tell. The others, however, did not have a national medium for spreading the word about this wild, beautiful territory.

Hog hunters on horseback

The exciting lives led by rough-cut Arkansas hunters and trappers provided fodder for many nineteenth-century periodicals. This artwork by Gilbert Gaul appeared in *Harper's Weekly*. It exemplifies the image many people had of early Arkansas.

Arkansas History Commission.

Passenger pigeons illustration from *Historic Arkansas*

In the early 1800s, passenger pigeons probably were the most common birds in Arkansas, numbering in the millions, perhaps billions. By 1914, the species was extinct throughout its range, a victim of unregulated hunting.

Arkansas History Commission.

For instance, "Uncle Mel" Robinson came to Jackson County in 1831 and lived later at Newport. He reminisced to a newspaper reporter that, "We could raise no hogs. The bears carried them off. One night a bear came and stole a pig and the next day we got the bear and the pig, too. The bear had eaten a little of the pig's shoulder and was asleep."

"We lived on bear, venison and possum," Robinson continued. "Any hollow log had a possum in it. I used to build fires around the cornfield to keep the coons from eating the roasting ears. The day I got married, I went hunting and killed three bears for the wedding, the first of December 1853" ("Chips and Shavings" 1952).

During territorial days, passenger pigeons were numerous and easy to kill. The *Arkansas Gazette* reported on October 27, 1841, that two men went to Fourche Bar just below Little Rock and killed nine hundred pigeons with fifteen shots. This may not be exaggeration. Glenn Martel, the Columbia County historian, studied the origins of place names in his area and reported in 1952 that the great flocks of passenger pigeons gave a creek its name. "There are those living here who remember their parents telling them how the wild pigeons roosted in the trees on Pigeon Roost Creek in such numbers that large branches were broken off hardwood trees" (Martel 1952). The *Van Buren Intelligencer* reported on June 21, 1845, "The wild pigeons have this year accumulated about Jackson [County] in this State, laid, and hatched out their young, and from a correspondent we learn are destroying the wheat crop to an alarming extent" ("Chips and Shavings" 1952).

Henry A. Hawkins first settled in Arkansas along the Red River about the time the territory was becoming a state in 1836. The fur and hide business was still a money-making industry in southwest Arkansas, and the woods were still full of bears, mink, foxes, raccoons, beavers, and otters. Trappers would go up the Red River as far as hunting was good, then build a boat or raft and float down the river, hunting and buying furs from other trappers and from trading stations along the river, to New Orleans where they would sell their furs. They would trap back up the river,

Panther art from *Big Bear's Adventures and Travels*

Florida panthers, or mountain lions, were among the animals most feared by early Arkansans. This illustration from *Big Bear's Adventures and Travels* was accompanied by this copy: "I throw'd back my gun to gin it to her, as she come; the lick I aimed at her struck across the shoulders and back, without doing any harm, and she had me."

Arkansas History Commission.

get home the best way they could and start down the river again (McIver 1955).

William Franklin Avera was brought to Camden by his parents in 1848. He recalled in his memoirs (edited by Henry Cathey and published in the *Arkansas Historical Quarterly,* summer 1963) that while timber rafting on the Ouachita River,

> every morning I would take a man to paddle the boat and go out in the woods and kill six or eight squirrels. It is a post oak country. The water stood up on the trees 10 or 12 feet. The squirrels were very fat and the best eating I ever saw. I would shoot one and the man would paddle up rapidly and I would pick it up. They were nearly all fox squirrels. I killed 76 while in the overflow. I cooked them myself every day and none of us ever tired of them.

Federal laws now protect the southern bald eagle and the Florida panther. In the early 1800s, however, they were considered fair game for shooters. The *Des Arc Citizen* reported on December 4, 1858, that John E. Ellis of Prairie County had killed an eagle near Hickory Plains. The eagle measured seven feet from tip to tip of the wings, and its spread foot measured eight inches across. In late fall of 1838, a widow McBride of Monroe County was told by one of her children that the family dogs had treed a panther within a half mile of her house. Mrs. McBride obtained powder and lead from a neighbor, molded bullets, and shot and killed the panther. The sound of the gun started up another panther nearby. It ran up a tree within a half mile of the other, and she killed it, too ("Chips and Shavings" 1952). Panthers were still plentiful in the late 1800s, at least in Mississippi County.

Now one of the premier agricultural counties in the United States, Mississippi County was "in 1902 a very sparsely settled area, about 90 percent of which was called a hopeless malaria ridden swamp . . ." It was then

> a hunter's and fisherman's paradise teeming with deer, bear, a variety of furbearing animals, wild turkey, and all kinds of waterfowl. A plethora of many varieties of fish existed. There were specimens of the alligator Gar up to more than six feet in length and weighing 200 pounds and more. The writer, on his first coon hunt had gone less than a quarter of a mile from camp when he and his guide found four raccoons up a persimmon tree.
>
> There were also predatory animals, particularly bear, panthers, and almost innumerable bobcats and catamounts, and packs of large timber wolves, all of which inflicted heavy losses on the settlers by destroying hogs, cattle and colts. In desperation, several of us secured a fine pack of wolf hounds from Missouri and Oklahoma with which we killed many wolves and catamounts. We captured some catamounts in hollow trees and logs. I kept several in a cage in Luxora until I was thoroughly tired of them. (Simonson 1947)

Herman Davis was Arkansas's greatest hero of World War I, according to historian Margaret Smith Ross. He was born three miles from Manila, then known as Big Lake Island, in Mississippi County. Herman and his father, Jeff Davis, spent most of their time hunting and fishing to supplement their meager farming income. The wild character of Herman's early environment can be illustrated by the fact that it was not uncommon for panthers to get on the roof of the Davis home. When this happened, Mrs. Davis and her small children would calmly stuff blankets into the fireplace to prevent the big cat from gaining entry and wait for the elder Davis to come home and kill it.

The tales Judge William F. Pope told, in *Early Days in Arkansas,* about his days as an early settler in Arkansas are among the more entertaining and historically significant ones. That he had a great appreciation for nature was illustrated by his recollection of stopping at Montgomery Point, one of the oldest and most widely known landings on the Mississippi River, while en route to Arkansas Territory. It was on Montgomery Point, he said, "that my ears were for the first time saluted by the ravishing strains of the mocking bird—that glorious mocker of the woods. It seemed to me that every tree and bush and shrub was vocal with the ever-changing notes of this wonderful bird of song."

Judge Pope tells another story about how the famed Bowie knife or "Razorback Toothpick" came into being. His story differs from the one popular now. The judge said James Bowie's brother, Reason Bowie, was a wealthy planter who lived at Walnut Hills in Lafayette County. Reason Bowie, Judge Pope said, was "a keen lover of the chase and spent most of his time in hunting the bear and deer, in which the country then abounded." While on one of his visits to Washington in Hempstead County, Reason Bowie called on blacksmith James Black and commissioned him to make a hunting knife of a design that Bowie had worked out. Bowie whittled the exact shape of the knife he wanted from the top of a cigar box. He explained to the blacksmith he wanted a knife that would disjoint the bones of a bear or deer without gapping or

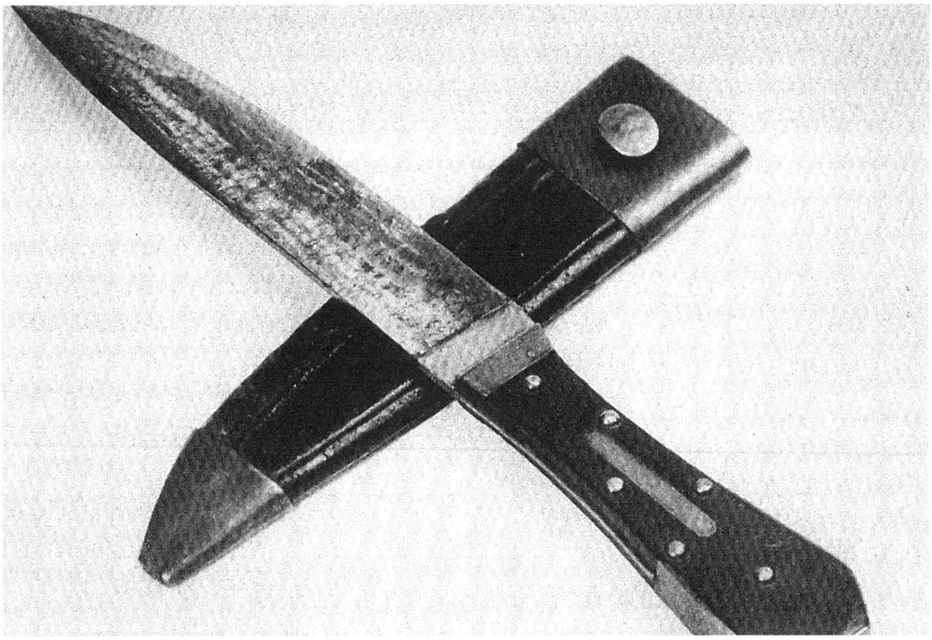

Bowie knife

A Bowie-type knife made by Arkansan James Black, creator of the famed Bowie knife.

Arkansas History Commission.

turning the edge of the blade. Black took the job on and turned out the instrument that became known as the Bowie knife. Judge Pope insisted the knife "was originated for use on wild beasts, and not on living man."

Judge Pope also recalled that in the early fall of 1835, Davy Crockett and a party of Tennesseans passed through Little Rock on the way to Texas to assist in the struggle against Mexico. When it became known that Crockett was in town, a committee went to invite him to a complimentary dinner. When the committee, chaired by Col. Robertson Childers, reached the hotel, they found several strangers sitting on the porch. On inquiring about Colonel Crockett, the committee was informed he was in the back yard, skinning a deer he killed about five miles outside of the city. They found the future hero of the Alamo in the back yard, "engaged in his bloody task . . . Referring to the deer he was engaged in skinning, he remarked: 'I made him turn ends at two hundred yards.' The Rifle Club of Philadelphia had recently presented Colonel Crockett with a magnificent rifle of the most approved pattern, the stock and barrel of which was richly inlaid with silver. [But] . . . the rifle with which he had made the deer 'turn ends at two hundred yards' was a long,

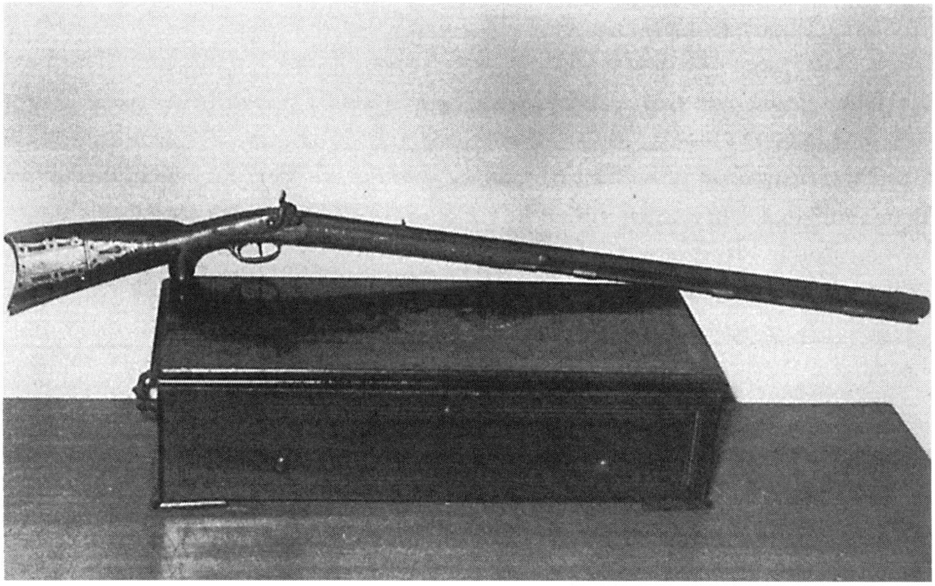

Davy Crockett's rifle

Rifle owned by Col. David Crockett. Crockett visited Little Rock in 1835.

Arkansas History Commission.

old-fashioned rifle, with flint lock, and which he tenderly referred to as 'Old Bet.'"

The hunting weapons used by most Arkansas pioneers were muzzle-loading, cap-and-ball rifles. For ammunition, settlers bought black powder and small bars of lead. They made bullets by shaving lead from the bars, melting it in a large spoon and pouring it into bullet molds.

J. J. Propps, formerly of Howard County, said in his 1967 writings that these muzzle-loading long rifles "could kill a deer at two or three hundred yards. A skilled marksman could fire and reload within a matter of seconds, much quicker than it takes for me to describe the process. We later used breech-loading shotguns with shells, but we reloaded our own shells. This was done by packing the empty shell with powder and shot and replacing the cap in the brass end of the shell."

Propps described his area of Howard County as a "deer hunter's paradise" that attracted many hunters. His own father "was always ready to hunt, and it vexed Mama that he spent so much time in the woods. One of my clearest memories about his clothing was that he always wore out the shoulder of his coat where he carried his hunting rifle," and he was "in the habit of taking his rifle with him even when he was going only a short distance from the house."

Thomas "Uncle France" Varner lived in Mississippi County near the Missouri border. When he was a boy, he took an old muzzleloader one spring morning in "turkey gobbling time" and made his way through bush and darkness until, about a mile and a half from his home, he located a gobbler in a tall tree. Varner quietly sat by a tree to wait for enough light to get a good aim, for there would be no second shot if the first one missed. As Varner sat there alone in silence, and while the proud gobbler was filling the whole forest with its calls, his attention was attracted by a noise at the end of a long fallen tree, the butt and stump of which were within a few feet of him. "Scratch! Scratch! came from four objects trotting into view on top of the log. The foremost one sprang to the stump and all were instantly seated; their heads were lifted skyward, and four wolves never created a more hideous rabble, especially to a lone and scared boy."

Varner pulled his muzzleloader down on the quartet's leader and pulled the trigger. "He tarried not nor stayed he" to see if anything vacated its seat on the stump, but with gun in one hand and hat in the other, he trod down vines, briars, and bushes as he made a hasty departure for home. When about halfway home, he met two of his brothers who had heard the wolves and the shot. By that time, daylight was coming. They went to the point and found the lead wolf "plum dead." The others and the gobbler had departed (Buffaloe 1961).

4 Early Days of Statehood

President Andrew Jackson signed the bill admitting Arkansas to the United States on June 15, 1836. In light of Arkansas's reputation, it seems appropriate that one of its most ardent sportsmen, C. F. M. Noland of Batesville (alias Pete Whetstone), who also was a lukewarm Whig as well as an editor, writer, humorist, lawyer, and politician, was selected to deliver Arkansas's new constitution to Washington for the use of Congress in adopting the statehood bill.

The rechartering of state banks was one of the most hotly contested issues in the General Assembly session of 1837. It also happened that a member from Marion County had introduced a bill, said to have been written by William F. Hicks of Little Rock because the Marion member was unable to write or at least unable to formulate a bill, to increase the bounty on wolf scalps to $2.50. On the second reading of this bill, an amendment was offered to the effect that the person killing a wolf should be allowed to deposit the scalp with the nearest justice of the peace, who would issue a certificate for it that would be receivable in payment of county taxes. J. J. Anthony, a member from Randolph County who had taken a leading part in opposing the rechartering of the banks, made an amendment to the amendment that the certificate issued by the magistrate also should be countersigned by the president of the Real Estate Bank. He was referring to John Wilson, the speaker of the house.

> This was a home thrust, which the Speaker very keenly felt, and he excitedly and peremptorily ruled Anthony out of order and commanded him to take his seat.
>
> This Anthony refused to do, saying that he stood there as the representative of the people of Randolph County and that he was entitled to the floor and should refuse to resign it at the dictates of John Wilson . . . The Speaker, livid with rage, left the chair, saying, "Then, I will make you." As he descended from the Speaker's stand, he drew his knife and advanced down the aisle to where Anthony was standing behind his desk. As Wilson approached nearer, Anthony, who was a very large and powerfully built man, over six feet tall, weighing 200 pounds, and by no means a coward said, in rather conciliatory tones, "My remarks were intended as a joke."

> Wilson, however, continued to advance, and as he did so, Anthony drew his knife, stepped into the aisle and awaited his antagonist's coming.
>
> Grandison D. Royston, a member from Hempstead County, who occupied a seat near Anthony's, raised a chair and thrust it between the now infuriated foes. Both men grasped the chair with their left hand and began cutting at each other, and Anthony inflicted a slight wound on Wilson's left arm. At the second attempt to reach his enemy, Anthony's knife flew out of his hand and he was placed at the mercy of his antagonist. Wilson raised the chair, and, stooping down to make his arm sure, plunged his weapon into his foe's abdomen. Anthony fell, exclaiming, "I'm a dead man," and immediately expired. (Pope 1895)

It was not an auspicious beginning for a new state.

This also was the year that Arkansas began to achieve an international reputation—though not for this incident, as worthy of notoriety as it might have been. Frederick Gerstaecker, German adventurer, arrived in Arkansas in 1837 for a visit that lasted until 1842. Through a series of letters to his family, published as *Wild Sports in the Far West,* Gerstaecker left a record of his travels in Arkansas that vied with James Fenimore Cooper's in depicting for Europeans "the pleasures of free life on the frontier." One of the sports described was bear hunting, a pursuit Gerstaecker appeared to enjoy with great relish and about which he wrote with enthusiasm. This is an example of what Gerstaecker told his family:

> At this season of the year our favorite sport was hunting. In the daytime we were busy cultivating our corn crop, but when night came we went into the woods and hunted in this fashion: A board about four feet long was fastened to the handle of an ordinary frying pan and then placed on the left shoulder. The fact that the board is flat lessens the chance of the frying pan's turning over. The pan itself is filled with small pine knots which when lighted make a bright, tall

Nineteenth-century bear and deer hunters

Frederick Gerstaecker, a German adventurer, traveled through Arkansas from 1837 to 1842. His record of that journey, *Wild Sports in the Far West,* draws an interesting picture of frontier life. Of Francisville on the St. Francis River he said, "The neighbourhood was composed of a wild, careless, and independent people of squatters and hunters, who rarely had a cent of ready money, but often brought in the produce of the forest, such as bear's fat, honey, smoked deer-hams, skins, and bear-hides, which they exchanged for powder and shot, whisky, and the other luxuries of forest life."

Arkansas Game and Fish Commission.

> flame. With the board across the left shoulder and the rifle in the right hand everything is ready for the hunter to start out. Since he must have a supply of pine knots available at all times, he must, if he goes alone, carry the stock of fuel and this is usually carried in a sack thrown over one shoulder.
>
> It may be easily seen that a hunter so encumbered will not have his left hand free to hold the rifle barrel when aiming at game. To remove such a difficulty a wooden fork is carved out and attached to the front part of the board. When game is spied, the rifle barrel is placed in the fork, and holding the flaming pan so that his head remains between the flame and the eyes of the animal, the hunter is in position to aim directly at a point between the buck's eyes, or if close enough to see directly the outline of its body, he can get a shoulder-shot. Once both glowing eyes of the animal are seen, the hunter needs only a steady hand to be certain of his game.
>
> The hunter, wise to the ways of deer and knowing that they come to the salt licks at night, goes to a "stand" and awaits the approach of the animal. The structure known as a "stand" is quite simple. Four poles, about five feet apart, are driven into the ground to form a square. A roof or top is next built by laying cross poles very close together. Moss and leaves are placed on the improvised roof and they in turn are covered by a four or five-inch layer of sand or dirt. The stand is usually built at a distance of from twenty to thirty feet from the lick itself. At night the hunter builds a bright fire of pine knots on the roof and protected by the layer of sand or dirt, he quietly waits within the stand. Although he himself cannot be seen, he is able to see everything moving within a radium of seventy to eighty feet. Unconcerned about the fire, a buck, stepping majestically, may approach the salt lick in the silence of night, only to fall dead when hit by a bullet of the hunter crouching in utter darkness not more than twenty feet away.

In another published work, *Western Lands and Western Waters,* published in 1864, Gerstaecker was purporting to describe the town of Francisville on the banks of the St. Francis River in 1852. He wrote,

> Wolves, bears, and panthers frequently at night imprinted their footmarks on these chief streets, and the next morning mine host of the City Hotel, leaving the business in the care of his wife and a small Negro boy, would set off with a pack of dogs in chase of his uninvited guests, frequently the only ones, though, he had. The neighborhood was composed of a wild, careless, and independent people of squatters and hunters, who rarely had a cent of ready money, but often brought in the produce of the forest, such as bear's fat, honey, smoked deer-hams, skins, and bear-hides, which they exchanged for powder and shot, whisky, and the other luxuries of forest life.

Overall, the picture presented by Gerstaecker depicted the state of the "downstream people" (Arkansas) as a wilderness filled with adventure and populated by rugged, hard-drinking, somewhat lazy, and often violent backwoodsmen who reveled in the sport of hunting. Critics differ about Gerstaecker's originality and accuracy. Regardless, insofar as the state's image was concerned, Arkansas was indeed a hunter's paradise—at least in the minds of his readers who had never been there.

A few herds of buffaloes were still in the Delta region when Gerstaecker visited. He accompanied a hunting party to the Bayou DeView and Cache River bottoms. After three days of fruitless searching, they found a herd of sixteen. A cow and calf

Left, **Alligator attacking man**

Right, **Wolves attacking man**

Illustrations from Gerstaecker's *Wild Sports in the Far West.* Gerstaecker depicted Arkansas as a wilderness full of adventure and populated by rugged, hard-drinking, somewhat lazy, and often violent backwoodsmen.

Arkansas History Commission.

were in the rear. Hoping to take the calf alive, they all fired at the cow; she fell, but the calf escaped. The taste of buffalo meat proved to be quite disappointing to Gerstaecker (Holder 1951).

Dr. Michael Dougan of Arkansas State University, Jonesboro, a former president of the Arkansas Historical Association and an authority on Arkansas in the Civil War, knows of no literature containing references to wildlife in the state during the war. It was agreed it would be logical to assume, however, that when military rations were short, as they often were, and the livestock of civilians had been decimated by previous raids or by the needs of the homesteaders themselves, both Union and Confederate troops turned to the land and its game to meet their needs during the Civil War.

There is evidence Arkansas remained relatively rich in game and fish resources through and after the war, for Capt. J. E. Lindsay in Dardanelle wrote a letter on March 16, 1866, to Maj. N. B. "Poly" Eison of Jonesville, South Carolina (given in *Arkansas Historical Quarterly,* spring 1969) in which he said,

> Poly, you ought to be here to hunt with me. There's all kind of game here. Deer in droves, the woods is full of turkey, and in the mountains but a few [miles] off, there's bears and panthers. There is game here of every description from a skunk to a wild Buffalo. I made a terrible raid on wild cats yesterday out at the Magazine mountain. (Reed 1969)

Further evidence of this is in a dispatch home from Elihu C. Backham, a soldier in Company K, 21st Arkansas Regiment, Confederate Army, whose unit boarded the steamboat *S. H. Tucker* at Pocahontas on February 14, 1862. He wrote,

Arkansas remained rich in game and fish resources throughout and after the Civil War. Capt. J. E. Lindsay in Dardanelle wrote a letter on March 16, 1866, to Maj. N. B. "Poly" Eison of Jonesville, South Carolina, in which he said, "Poly, you ought to be here to hunt with me. Theres all kind of game here. Deer in droves, the woods is full of turkey, and in the mountains but a few miles off, theres Bear and panthers."

Arkansas Game and Fish Commission.

Hunting photo of men with deer, bear and, turkeys

Hunters with wild turkeys

Turkey hunters

> We landed at Jacksonport on the 15th, and camped on the bank of the White River. The next day we started up the Black River. I remember while running up the river there was a gang of wild turkeys in the timber and we were firing on them at long range with our old banded muskets and David Bone killed a fine one. Just as the boat sidled against the bank, he (Bone) jumped ashore, and got the turkey, and was back on board before the boat left the bank.

What the Civil War undoubtedly did, however, was expose Arkansas, including its abundant game, to outsiders and to the already existing, but rapaciously developing, ethic that there were no limits to any natural resource that America offered, be it land, minerals, timber, wildlife—and no limits to the opportunities to exploit them. Arkansas was not immune from the exploitation ethic as the last three decades of the nineteenth century demonstrate.

Dr. Albert A. Honor of Boston, Massachusetts, who grew up in Helena in Phillips County, is among those who innocently represented this ethic in a letter he wrote August 28, 1967. The elderly physician told the folks "back home" that virtually all the boys who grew up in Phillips County in the nineteenth century hoped to go on a bear hunt down in the Bottoms—some of them even dreamed of being one of those who killed a bear. Dr. Honor confessed that he never went on a bear hunt, but he had observed bear hunters as a child. His recollections were published in the *Phillips County Historical Quarterly,* September 1967:

> Some mornings in December, I would hear hounds baying and look across Porter Street to see 20 to 30 dogs in Mr. Opp's garden. These dogs had come by river boat from Kentucky. They were the property of some men who came annually before Christmas to go bear hunting with Mr. Opp. After one or at most two days the dogs would be gone. They were bound for "The Bottoms," and would hunt in the canebrakes—and big timber between Helena, Arkansas City, and the White River.

Early bear hunters

The "Bottoms" between Helena and Arkansas City was a popular bear hunting area in nineteenth-century Arkansas.

Arkansas Game and Fish Commission.

> There was no intention of shooting any bear; this was below the dignity of a bear hunter. The hunting was done by dogs who would find, chase, and "bay" a bear. (Baying meant to encircle the bear by the dogs—making its escape impossible.) Once the bear was bayed he would fight the dogs that ran at him from every point of the bay circle.
>
> When the men reached the bay, one or more of them would go in to kill the bear with a knife. Mr. Opp had a great reputation as a killer of bears. Before going in to kill the bear, Mr. Opp would exact a promise of no shooting unless "you think I am being killed."
>
> With every pack of bear hounds was one or more shepherd dogs. These dogs did not enter the chase nor the fight until their master went in, then they were fresh and kept biting the bear from unexpected quarters while their master was busy cutting the bear's throat.
>
> Once the bear was dead, it was lifted from the ground and hung by its fore feet, using rope or "galluses," or anything handy, to a limb five or more feet from the ground.
>
> Then the bear was dressed and the heart, lungs, liver, and other "innards" given the dogs, some distance away from the tree to which the bear was tied. While all the dogs feasted on bear "innards," the bear was skinned, untied, and taken back to camp. The first bear to be killed on the hunt was consumed by the hunters while in camp. There were many stories about bear stew and camp stews. Some of the stories were undoubtedly embellished by the whiskey drunk directly from stoneware gallon jugs which were an integral part of bear hunting equipment.
>
> Other bears killed were brought back to Helena and often one was displayed in front of Mr. Burton's meat market on Rightor between Cherry and Ohio Streets. There a choice cut could be purchased by a non-hunting game lover. At the same time, Mr. Burton might have wild geese, ducks, quail, or venison for sale.

George H. Robertson of Paducah, Kentucky, is one of those who came to hunt bears in what was "practically a wilderness" below Helena in the vicinity of Elaine. According to the article "Bear Hunting in Phillips County," published in the *Phillips County Historical Quarterly,* he and his party would load their hunting gear on the packets *Jim Lee* and *Kate Adams* and land near Modoc. "I remember on one occasion, it took 10 wagons and 40 mules to move our camping equipment into the woods where we pitched camp," he said. Often there would be as many as thirty men in the hunting party. And how many bears would they kill on a single trip? "Well," he replied, "I remember that on one of them we killed 32."

Most cities of any size had game markets in the late 1800s. For example, several markets were between Louisiana and Center Streets on Capitol Avenue in Little Rock at which buyers could find deer, bears, wild turkeys, geese, ducks, quail, and prairie chickens. Most Little Rock restaurants and hotels served game. An article in the *St. Louis Globe-Democrat* on January 23, 1916, by E. T. Gether, then the outdoors editor, attests to the truthfulness of this and to how long the exploitation ethic prevailed. He wrote,

> Rabbits, coons and opossums are still the hunters' reliance for sport in this locality. There seems to be no end to the rabbit supply this season, judging from the quantities received by St. Louis commission dealers. Landauer, the Third

Mena meat market

Game markets were found in most large Arkansas towns in the late 1800s. This photo shows the Ben C. Strauss market in Mena, circa 1905.

Arkansas Game and Fish Commission.

> Street rabbit king, says: "We have handled more rabbits than ever before. They come from Missouri, Arkansas, Kansas, Oklahoma, Texas and New Mexico. The smaller, light-colored rabbits come from the latter two States. The other states send the regular cottontail rabbits. We have handled from 5,000 to 10,000 rabbits in a single day, besides smaller quantities. They are shipped to Indiana, Ohio, and Pennsylvania, where rabbits are scarce and high-priced."
>
> To the person in this city who sees all those rabbits piled up at one time, it is a very forceful illustration of the wholesale manner in which game of other varieties was handled.

In January 1968, Harry Lee Williams wrote an article about the early days of this century when he lived in Jonesboro, fifteen miles east of the Cache River bottoms and fifteen miles west of the St. Francis River bottoms. He called it the "greatest game reserve on earth." In those years, he said, market hunters like John McDaniel, Ed Mattix, John Thomas, and Nick McCall from a few miles southwest of Jonesboro; Aaron Cole from eight miles west; and Frank Lane from ten miles northeast flooded the markets with game, shipping the excess to markets in Memphis and St. Louis. Dan Cantrell and others from Marked Tree and Frank Biggs of Oak Donnick shipped bears killed in canebrakes along the St. Francis River. Venison steak sold for five cents a pound, bear steaks for seven cents a pound, and wild turkeys for fifty cents each.

Williams recalled Frank Lane telling him that in the late 1890s, he averaged 300 deer a season, selling them to a merchant at Alicia (Lawrence County) for two dollars each. The merchant shipped them to St. Louis, receiving five dollars a deer. Williams also said that Charles Neeley, an early-day merchant at Fisher, told him that in the late 1890s, 2,000 deer had been shipped out of Fisher to St. Louis and Chicago the year before. While hunting on Big Creek in 1909, Williams said he met an old trapper named Kennedy, who claimed to have sold 160 deer the previous winter to steamboats and storeboats along the lower river.

"In those days," the *Arkansas Deer* reported in 1924 in the unattributed article "What Was Here Fifty Years Ago," "there were but few railroads in the State and all the transportation of the country was by steamboats, and it was nothing uncommon to see a steamboat go down the river with the forecastle strung all over with game of every kind, from a rabbit to a bear. Deer was so plentiful in the country that many farmers would put up smoked venison, enough to last the entire winter, and some of them used it in place of beef and pork."

Before the United States signed a waterfowl treaty with Canada in 1917, Williams said in his article, market hunters on Big Lake in northeast Arkansas earned an average of one hundred dollars a day selling ducks. They shipped their day's kill in ice-filled barrels to Memphis, Kansas City, St. Louis, and Chicago. When the treaty outlawed market hunting, Williams said market hunters became guides, taking duck hunters out on the lake for ten dollars a day.

Williams also said he had met many market fishermen while hunting on the lower White River and the Mississippi River. Jess Webster and George Griffin, for example, "caught thousands of pounds of fish every week in nets and sold them to fish buyers who visited them on Monday and Thursday." The Moss Fish Company of Clarendon handled more than one thousand tons of fish weekly. Williams quoted Webster as saying that in 1909, he made sixteen thousand dollars in six months the previous year selling buffalo at two cents a pound, catfish at three cents a pound, and crappie and bass at five cents a pound.

Some streams in Arkansas were more productive and could withstand the commercial fishing pressure better than others. The White certainly was one of these. The Game and Fish Commission recognized it in 1916 as one of Arkansas's, and probably one of the country's, greatest fishing streams. "The amount of fish annually caught on lower White River, even after years of continuous slaughter in that portion of the river, is something almost unbelievable," the commission asserted. In the twenty months preceding the commission's report, 126 carloads of fish averaging 24,000 pounds to the carload, caught on the lower White River, were shipped from Rosedale, Mississippi. In addition, there were fifteen hundred express shipments from Rosedale averaging three barrels to a shipment and 200 pounds to the barrel, a total of 9,000 pounds. This makes a total of 3,924,000 pounds of fish from the lower White that were shipped from Rosedale in twenty months. Equally as

William H. Lester fishing the White River at Batesville, 1895

Excellent sport fishing drew many anglers to the state's larger rivers, but most fish were caught by commercial anglers who sold them to fish buyers who then supplied them for sale to the public. Around the turn of the century, the Moss Fish Company of Clarendon handled more than one thousand tons of White River fish weekly.

Arkansas History Commission.

many pounds were shipped from points farther upriver. "From the State at large," the commission said, "we can safely estimate the shipment at 20 million pounds of fish in the last 20 months. The prices range from four to ten cents per pound. At a reasonable price of five cents per pound, this amounts to $1 million."

Most Arkansas streams did not fare as well, however. The Saline River was among these. Lance Hemingway recalled in 1924, "Over most of its winding course of a hundred miles or so the Saline is a beautiful stream (I know nothing of its reaches below Ozment's Bluff), and during my boyhood in Drew and Cleveland Counties it was plentifully supplied with fish of practically every kind found south of the Ohio River." "In those days one could go to the Saline with a bucket of live minnows, or a can of angleworms, and a 'fly' hook rigged for catching topwaters or shiners, and take enough black bass, striped bass (then known as 'barfish'), crappie, and various small sunfish, to fill a fair sized gunny sack. The same was true of the cut-off at Collins, the system of lakes near Comminto, Bayou Bartholomew, Williams' Lake, the Niemeyer and Dyer and other cypress brakes, and all the smaller creeks and 'spring branches' in the county."

Hemingway lamented, "If such fishing can be had in those waters . . . it is distinctively news to me, for when I visited them last, dynamite, the trammel net, and the long, small-mesh seine had stripped them of practically everything worth taking. Indifference on the part of the general public toward constant abuses of 'rights and privileges' had brought the inevitable result."

In his article cited earlier, Harry Lee Williams relayed another story that he said came from Jimmy Wallace, who originally lived at Carlisle. Wallace said men from St. Louis and Chicago would check into local hotels and bet on who could kill the most prairie chickens. The bets ran as high as one thousand dollars. "Disgustedly,

Jimmy told me that this mongrel breed would go out on the prairie, kill from 80 to 100 a day, pull their heads off to bring them back to count and throw the chicken away."

Bennie Frownfelter Burkett of Stuttgart remembers stories she heard about wildlife on the Grand Prairie. Her great-grandparents came to the Grand Prairie in 1883, settling in what became known as the community of Turley. The first of their four children, George Howard Frownfelter, was her grandfather. His son, Edward Benjamin, was born November 1, 1891, the youngest of five children, and died in 1984 on the Grand Prairie at the age of 93. Edward Benjamin was a year old when his father died, so the tyke was doing his part to help put food on the family table by hunting and trapping by the time he was nine years old. The first animal he trapped was a mink. The second was a skunk, and the next was a raccoon. "I remember getting a dime for a skunk hide," he told a Little Rock newspaper reporter Jan Meins in 1980. When he caught the raccoon, he had a dog with him but no weapon with which to kill the furbearer. "That dog jumped on him," Frownfelter recalled. "Between the dog and myself, we finally conquered the coon. It takes a good dog to whip a coon."

The buffaloes and beavers already were gone from the Grand Prairie by the time Edward Benjamin started hunting, usually with a carbide lamp strapped to his forehead. Ducks, geese, quail, and prairie chickens were thick. In fact, there were so many prairie chickens they "looked like a black cloud when they flew over," Frownfelter said. One afternoon of quail hunting would yield tow sacks full that became meals for the Frownfelter family and for the tenants on their farm. In summer, prairie chickens would fly over the picket fence around the Frownfelter home to use the water trough. When the prairie chickens sensed human presence, they would flee, and some caught their necks in the picket fence as they tried to dart underneath it. Edward Benjamin and his brothers would race along the fence, snatching up the prairie chickens so they would have fresh meat for lunch.

Burkett recalls her mother "frying up" pans of quail her father brought in from hunting. There were lots of blackbirds, too, and her mother made them into

Early duck hunters with mallards

Early Arkansas settlers found a land teeming with waterfowl and other wildlife. But by the turn of the century, many species were on the brink of extinction, victims of unregulated hunting and habitat destruction.

Arkansas Game and Fish Commission.

Prairie chicken

Prairie chickens were numerous on the Grand Prairie when the French settled Arkansas Post. After the Civil War, the birds were hunted for the commercial market and their population sharply declined. In 1893, the Arkansas legislature declared a closed season on prairie chickens for five years, but interest in protecting this species came too late. The development of rice culture on the Grand Prairie destroyed its habitat. The last Arkansas prairie chicken was killed in 1917.

Leonard Lee Rue III.

"delicious pies." But Burkett never saw a Grand Prairie prairie chicken herself; she was born too late for that, the last one being killed in 1917. Her father told her about groups of men from Little Rock and Memphis who would ride with their guns to the Grand Prairie in spring wagons. The men divided into clusters of two and three and vied to see which group could bring in the most prairie chickens. This was a "very popular sport" in the late 1800s and early 1900s, even though the 1889 General Assembly passed a law closing the season on prairie chickens from February to September of any year. Obviously, the law was enforced poorly, if at all; in addition, the legislature failed to set a bag limit on prairie chickens during the open season. The legislature went further in 1893 and banned hunting of prairie chickens for five years, but its interest in protecting the fowl came too late.

Unlike the prairie chicken, the buffalo left a legacy on Arkansas's prairies. James P. Neal in his article written about 1885 said that when he first became acquainted with the Fayetteville area in 1829, the buffalo already had receded some fifteen or twenty miles to the northwest and that when the settlers wanted buffalo meat, they took a wagon or a packhorse and went after it. Buffalo paths, however, were still numerous, leading primarily from one salt lick to another. These licks were scattered all over the prairie where Neal lived, perhaps one for every three acres of land. They were depressions, filled with water that was a "little brackish" in wet weather. In summer, they dried, and the earth permeated with salt. The buffaloes frequented them, hence the name, "buffalo licks." Burkett said that on the Grand Prairie, these depressions were called "buffalo wallows," and many reservoirs that today hold water for rice irrigation are located where the licks or wallows used to be.

Not all hunting was for consumption of the game, however. For example, J. J. Propps of Tulsa, Oklahoma, said one of the joys he had growing up near Galena in

Howard County was listening to fox chases in fall. His description of a fox chase from a young boy's perspective was published in "My Childhood and Youth in Arkansas" (*Arkansas Historical Quarterly,* winter 1967):

> People for miles away would come and bring their fox hounds and go together for fox hunting. Sometimes there would be 20 or 30 dogs. They would go into the game country north of our home. The hounds would hunt through the forest until they struck a trail where a fox had been. The trail might be cold, but the dogs would slowly hunt back and forth until the trail became warm. Then the dogs would make the woods ring with their barking while following the fresh trail. They could be heard for miles running first in one direction then in another. Some of the men would follow the dogs on horseback, but most of them would station themselves on some eminence from which they could hear in all directions. If the men on horses thought the dogs had a chance of catching the fox, they tried to follow and keep in hearing range. When the dogs were close enough on the trail, the fox would not have time to play along. Sometimes when he came to a rail fence, to throw the dogs off the trail, he would climb the fence along the top rail as far as he could or he might double back, jump off the fence and run back through the woods almost in the direction from which he came. Br'er Fox was very keen at outwitting the dogs and throwing them off his trail. But smart fox-hounds would move off in a circle trying to strike the trail again. The fox would run fast until the dogs were out of hearing. After awhile the dogs could be heard coming back following the fox on the new trail. Frequently, this would go on all night.

Raccoon hunters

Mr. Earnhart and Brad Hogan in Faulkner County, circa 1900.

Arkansas History Commission.

Propps apparently did not participate in the fox chases, but he hunted and trapped at an early age. "Sometimes we would catch a rabbit in a trap," he recalled, "but they often escaped by cutting the pieces of timber of which the trap was built." Propps confessed he was far more interested in quail trapping because it was the most delectable game he caught, and he "caught as many as half a dozen at one time." He regarded the squirrel as good eating, too, but they were taken with guns, not traps. Another interesting diversion for Propps and his companions was hunting raccoons and opossums at night:

> We would take our dogs into the woods where they would soon tree a raccoon or an opossum. We then cut the tree down, holding the dogs until it was nearly ready to fall . . . The thing we enjoyed most was a fight between the dogs and the raccoon.

Propps related how he trapped raccoons and mink in steel traps anchored to a stake and placed in their runways at the water's edge:

> We covered the traps with leaves and baited them with a bird of some kind that we had killed. The animals would have to walk around where the trap was set in trying to get the bird and would step into the traps. Because a mink or

a 'coon would gnaw his leg off to get free, we always set the trap in such a way that it would fall into the water and drown them.

Propps was a savvy hunter and trapper at an early age. His article, however, also reveals that the "killing for killing sake" ethic was to be found in young boys:

> On the farm as we were growing up from the age of 7 or 8 up to 20 or more, I hunted birds in many ways. First, perhaps, just roaming with two or three other boys . . . We hunted often; yet we had no guns to shoot anything we might want to kill. As is often true in growing boys, we were a group of little savages, trying to kill everything that came within range of our accurate throwing arms. We roamed up and down the little clear stream that ran back of our house the year around. We would stroll for several miles, seeking to kill anything we saw—birds, snakes, lizards, mosquito hawks, fish in shallow water, and even squirrels that might be so unwise as to go into a hole in a tree, or into a nest of leaves and sticks high up in a tree.

This ethic was not confined to Arkansas's countryside. According to the article "Passenger Pigions [*sic*] over Little Rock" in *Arkansas Game and Fish,* the *Arkansas Gazette* reported in August 1859 the city was full of disappointed boys because an unusually heavy rain the day before had delayed the arrival of passenger pigeons that flew over the town morning and evening in flocks of hundreds of thousands. With double- and single-barrel shotguns, parties of boys would station themselves at various points and fire simultaneously into the flocks passing over, bringing down hundreds of the birds. It was the rule of one party of boys that when the shooting was over, the pigeons killed would be placed in a pile encircled by the boys. One of the party would divide the pigeons equally to the members, placing one at a time in front of each, the odd one going to the boy making the division. The boys took the pigeons home, where they were made into pies. When some of the boys had more pigeons than their families could use, they took the surplus to a local bakery where they swapped them for hunks of ginger cake and glasses of sweet cider.

The youngsters learned their behavior patterns from adults, as a 1924 anecdote in "Father Teaches Conservation" in the *Arkansas Conservationist* illustrates: While looking at a black bass in a wildlife exhibit at the 1924 State Fair in Little Rock, a teenage girl was heard to remark "they had killed lots of those last summer." In response to questions from an adult who overheard her, the girl told of a vacation trip her family had taken through the Ozarks, and she boasted of her father's exploits shooting bass with a high-powered rifle. "The water was perfectly clear," she said, "and Daddy had no trouble seeing the nice ones."

This is not to suggest all of the wildlife killing, especially by young people in the 1800s and early 1900s, was wanton slaughter. To the contrary. Dee Simpson and his brother Bunt used their ingenuity as seven- and nine-year-olds to put food on their family's table. They had no money, so they scoured the countryside for discarded cast-iron cookstoves. When they found one, the brothers broke the brittle metal into chunks that became ammunition for their slingshots. Dee was adept enough he could "hit a 'chuck and roll him over" with the iron missile. Simpson won his first gun—a single-shot .22 rifle—and ten-cents worth of ammunition at age nine by selling twenty-five-cent jars of Cloverine salve.

5 Taming the Wilderness

It has been shown how, through the writings of C. F. M. Noland, Friedrich Gerstaecker, and others, Arkansas acquired its Bear State image. Opinion makers on the eastern seaboard, enamored with the romanticism of the frontier presented by James Fenimore Cooper and other earlier writers, easily accepted the description of the "downstream state" presented in Pete Whetstone's tales. Many Arkansans did not, however. They wore the mantle of the Bear State uneasily and never adopted it officially as Arkansas's popular name. Borrowing from Whetstone's adventures, Louisiana writer Thomas Bangs Thorpe wrote a story entitled "The Big Bear of Arkansas." Thorpe's bear, hunted by an unnamed resident of Shirt-Tail Bend, was a larger-than-life symbol for frontier society. According to historian C. Fred Williams, it helped seal Arkansas's rustic image in the eyes of others.

There was a westward movement in the United States at the time, so it is questionable whether Arkansas's increases in population can be attributed to any campaign to attract settlers. Certainly it did not hurt. In the 1840 census, the first one after Arkansas became a state, there were 97,574 people living in Arkansas. In ten years, that number shot up by 215 percent to 209,897 inhabitants. The population more than doubled again during the 1850s, climbing to 435,450. The growth rate slowed to 11.3 percent during the decade of the Civil War but accelerated again during the 1870s, reaching 65.6 percent and a total population of 802,525 by 1880. Arkansas topped the million mark in 1890. It took another ninety years that included two decades (the 1940s and 1950s) of population loss for Arkansas to reach two million residents.

Before the Civil War, travel was either on foot or by steamboat or wagon. But the spurt in Arkansas's growth during the 1870s and 1880s coincided closely with the construction of railroads. In addition to transporting settlers, the "iron horse" made it possible for the timber industry to flourish. During the 1800s, the timber industry moved from New England to the Middle Atlantic and Great Lake states, finally branching out to the South and the Pacific Northwest. As the timber in one region was cut out, the industry moved on. Lumbermen moved into the South in two waves, beginning in the 1880s. One group migrated from New England and the Middle Atlantic states, the other from cut-over pinelands in the Great Lake states. New companies were formed, or old companies simply moved south. They bought

BALLOU'S PICTORIAL

M. M. BALLOU, | CORNER OF TREMONT AND BROMFIELD STS. BOSTON, SATURDAY, APRIL 7, 1855. $3.00 PER ANNUM. 6 CENTS SINGLE. | Vol. VIII., No. 14.—Whole No. 190

ARKANSAS.

Arkansas is, as our engraving beneath represents, a frontier State. The wild Indian yet occupies her borders, and emigrant trains are seen traversing her prairies with their attendant hunters and guards. A story, based upon the adventures of the train seen in the engraving, will be found on page 211. But although Arkansas is one of the "new States," not having been admitted into the Union until 1836, her territory was visited at an early period of our national history by European explorers. The gallant De Soto, with his Spanish gold hunters, and the energetic La Salle, with his French traders, visited Arkansas. French colonies sprang up along the river Mississippi, but the interior was not colonized until the older States sent forth their stalwart pioneers. When the last census was taken, the population of Arkansas was 209,897. Arkansas has 362 churches, mostly of the Baptist and Methodist persuasions. In 1840 she had no college, 8 academies and 113 primary schools, but in 1850 she could boast 3 colleges, 90 academies, and 353 primary schools, showing that her citizens are aware of the necessity of education. She has nine weekly newspapers and three published semi-weekly. Agriculture is the prominent occupation in Arkansas. Her live stock, in 1850, was valued at $6,647,969; and she produced 8,893,939 bushels of Indian corn; 218,936 pounds of tobacco; 65,344 bales of cotton, weighing 400 pounds each; nearly one million bushels of potatoes, and grains in like proportion. Over two thirds of her male population are farmers, and the census shows but one banker, one bookbinder and one musician. There were 448 physicians, 724 lawyers, 165 saddlers, 84 printers and 4 editors. Literature has not as yet made much progress in Arkansas; but she can boast of Albert Pike, who, though not born within her borders, yet has identified himself with her progress. His Hymns to the gods of Greece have met with high favor in Europe, and he has won equal fame as a poet, a soldier, a lawyer and a true friend of all good fellows. The future destiny of Arkansas, it is safe to predict, will be a brilliant one. Her forests are fast falling before the sturdy settlers, and soon the unbroken wilderness will be reclaimed. "Westward the star of empire takes its way;" and Arkansas, ere the close of the present century, will be the very heart of our republic.

Ballou's Pictorial

During the mid-1800s, many eastern newspapers romanticized life on the Arkansas frontier.

Arkansas History Commission.

great tracts of timberland on which they built mills and established towns for the workers and railroads to haul logs to mills and lumber to market (Clepper 1971).

A small sawmill began operating in Helena in 1826, ten years before statehood. It was the first in the Delta and the earliest facility of its kind reported in Arkansas. Large-scale lumbering got underway in the Delta in the 1870s, and some companies bought enormous amounts of land there solely for the timber (Holder 1951).

Logging in the Ozarks began by 1879, although there were only about ten steam sawmills operating in the area. Logging in the Coastal Plain and the Ouachitas began in the same period. Few people in the nineteenth century thought the timber wealth of the United States, or of Arkansas, would be exhausted. At first, timberlands in Arkansas were "high-graded," a practice of taking only the biggest, best, and most accessible trees. Later, some areas practically were stripped, entire watersheds almost denuded. By the end of the 1800s, choice timber species in the Ozarks, such as cherry and walnut, were hard to find. Virgin white oak and pine remained only in the most inaccessible locations (Bass 1981). It was believed at the time that all the cleared land would be needed for farms, and what cheaper way of doing it than by logging and uncontrolled forest fires? The problem was that lands with the most marketable timber—the Ozarks, Ouachitas, and Gulf Coastal Plain, for instance—were too sandy or rocky for profitable agriculture (Stoddard 1968).

When fire followed logging operations, as it frequently did, young timber was destroyed, delaying renewal of the timber resource. The settlers and mountaineers considered fire a time-honored farming practice. Nuttall observed the practice in his 1819 travels in Arkansas. He wrote in his journal about walking into a prairie near Fort Smith and finding it as undulating as the nearby woodlands, but Nuttall could find no reason for the absence of trees, except for the "annual conflagration." Homesteaders not only used fire to clear land but also to kill insects and to open the woods for grazing and hunting. In the process, fire damaged many trees as it burned uncontrolled over large areas each year. Rains then washed away exposed topsoil,

Floating logs on the St. Francis River at Madison

Lumbering was somewhat limited in the mid-1800s because businesses like the Wheeler and Company Sawmill in Madison (St. Francis County), pictured here, had to rely on steamboats, wagons, or rivers to carry logs to sawmills.

Arkansas History Commission.

Log train

Large-scale lumbering began in Arkansas in the 1870s. Railroad construction during this period provided a better means for hauling logs to mills and lumber to markets.

Arkansas History Commission.

leaving the eroded land poorer and silting downstream waters. Floods also became more frequent.

In the Delta, agriculture expanded slowly at first because of periodic river overflows. Farmers also had to contend with exceedingly poor drainage. The St. Francis Levee District was created in 1893, marking the beginning of orderly and sustained flood protection and reclamation of overflow lands (Holder 1951). This district embraced what once was the worst part of the "Great Swamp." Other levee districts soon began organizing throughout the Delta.

After lumber companies stripped the region of its best timber (and sometimes all of it), they sold the land at cheap prices because the concept of reforestation had not occurred to most of them. Marked expansion in agriculture followed. Levees were built along the reaches of the Mississippi River and other major streams most prone to flooding, and drainage canals were dug. Over most of the Delta, reclaimed land went into cotton production. The area east of Crowley's Ridge quickly became one of the outstanding producers of cotton in the world. After rice successfully was grown on the Grand Prairie, land that failed when planted in other crops was soon given over to rice. More than half the Delta—445,000 acres or 56 percent of it—was improved through drainage by 1926.

Some say the lack of vision by lumber companies is at least "understandable" when it is remembered the trees cut were 150 to 250 years of age. "Even the shrewdest businessmen in 1917 could hardly have been expected to invest capital in reforesting and protecting land whose next crop—so they thought—would not be ready for 100 years or more" (Clepper 1971).

Fire in the woods

Arkansas homesteaders' cabin

Arkansas homesteaders used fire to clear land, kill insects, and open the woods for grazing, hunting, and home building. In later years, however, fire was considered a threat to timber holdings, and Arkansas lumbermen began looking for ways to protect the state's forests.

Arkansas Game and Fish Commission.

Because of the combined effects of overhunting, the population increase, the cut-out-and-get-out philosophy of the timber industry and the flood protection–drainage work going on in the Delta, Arkansas had taken on a peculiar silence by 1915. Few deer remained. Wild turkeys and bears were the stuff of mountain dreams and campfire stories. Legend became the final habitat of the wolf, so thoroughly had parts of the state been hunted and trapped (Bass 1981).

Although few recognized it, some seeds of hope for the future were being sown during the late 1800s and early 1900s. The earliest laws passed by Congress affecting the country's forests, between 1799 and 1831, were intended to insure supplies of live oak for shipbuilding. During the nineteenth century, several states became interested in protecting their forest resources, first with special boards to study the forest situation, then with attempts to encourage tree-planting bounties, neither of

Hunters with deer fawn

By the late 1800s, Arkansans were beginning to see the effects of years of unregulated hunting and habitat destruction. Many species of game and fish were disappearing at an alarming rate, yet people in the state clung to the belief that government had no business tampering with their "God-given rights" to hunt and fish as they saw fit.

Arkansas Game and Fish Commission.

which was very successful. In 1875, the American Forestry Association was organized to educate the public about the need for conservation measures. In August 1876, Congress authorized two thousand dollars for the first federal office devoted to forestry, and Dr. Franklin B. Hough, a physician, became its first agent. His task was to gather data on forests and forest products, European forestry practices, and on the means to preserve and renew forests.

By 1891, it was apparent to many that forests represented a great but vulnerable national asset that for the sake of posterity should be protected from unbridled exploitation. That was the year Congress authorized the withdrawal of forest reserves from the public domain, to be administered by the Department of the Interior. President Benjamin Harrison set aside the Yellowstone Timberland Reserve (now part of Shoshone and Teton National Forests) in 1891. By the end of his term, Harrison had set aside 12.4 million acres of forest land to protect it from speculation. His successor, Grover Cleveland, added 25.9 million acres, and in 1897, Congress enacted legislation to insure proper care, protection, and management of public forests; to authorize hiring guardians for the reserves; and to open them for regulated use.

Some do not give the federal government as much credit as it gives itself for these developments. Although certain government officials and members of Congress were aware of the resource (habitat) destruction going on around them, the government itself was not the motivating force in conservation, according to author Henry Clepper. In truth, he said, "The indifference of public officials in high places created an enormous inertia difficult to overcome. How could it have been otherwise in an era when the logger's double-bit ax and the crosscut saw were symbols of progress?"

It would be a gross distortion of history to suggest all lumbermen and large forest owners were indifferent to what was happening to forests in Arkansas and the rest of the South. Many were aware of a social and economic need to restore the resource. Because wildfire was the principal natural hazard to timber ownership, owners naturally sought ways to reduce or eliminate the danger. Early in the 1900s, several companies in scattered locations throughout the South tried organizing forest protection forces for detecting and suppressing wildfire. But individually they could do little, especially in the absence of any public agency for forest protection (Clepper 1971). With passage of the Weeks Law in 1911, states were encouraged to enter cooperative forest-fire protection arrangements with the Forest Service. Arkansas, however, did not establish a forestry commission for this purpose until the 1930s.

The Crossett Lumber Company also is given credit in timber industry histories for enlightened management practices during the critical period of the 1890s and early 1900s. The Arkansas company, which built a sawmill at Crossett in 1899, sought technical advice early. A report on its land holdings was prepared by the 1912 class at Yale University School of Forestry. An estimate of mature timber indicated at the existing rate of cutting—more than 200 million board feet annually—harvesting could continue for eighteen or twenty years. The report said, "The land cannot be kept permanently in forest by the present company, and it is planned ultimately to dispose of it as farm lands." The stands were in unsatisfactory condition; severe fires had burned through the slash, destroying immature pole timber. Worthless oaks and scrub sweetgum were coming up. A Yale professor asserted, however, it was possible with proper management to maintain a perpetual supply of timber on Crossett's holdings. The company heeded the advice and hired one of the country's new professional foresters to assist it.

Once conservation became part of federal government policy, implementation moved forward. On March 14, 1903, President Theodore Roosevelt issued an executive order, seemingly insignificant at the time, that made the federal government responsible for providing refuge for the brown pelican. The order set aside Pelican Island, a two and one-half–acre tract on the east coast of Florida, as a bird sanctuary. It was the first national wildlife refuge and, in effect, established the land management responsibility of what would later become the U.S. Fish and Wildlife Service of the Department of the Interior.

Big Lake in Mississippi County was Arkansas's first national wildlife refuge. Established in 1915, it is one of the oldest refuges in the country. Covering almost seven thousand acres and composed of an intricate maze of lakes and creeks, Big Lake essentially is a cypress-tupelo swamp with a small amount of upland. Located near Blytheville, twenty miles west of the Mississippi River, the refuge surrounds the body of water for which it was named and attracts a wide variety of waterfowl and other birds.

Following establishment of forest reserves in the West, demands for similar reserves were made for the East and South. Arkansas was one of few eastern and southern states with extensive areas of public domain at the turn of the century. On December 8, 1907, Roosevelt signed a proclamation establishing the Arkansas National Forest in the Ouachitas south of the Arkansas River. (In April 1926, the area was renamed Ouachita National Forest in recognition of the fact that it extended into part of Oklahoma.)

Arkansans reading their newspapers on Sunday, March 8, 1908, learned that two days before, the president had signed an executive order establishing the Ozark National Forest, making it the first protected hardwood forest in the country.

Later experiments on these two national forests helped point the way toward the return of a resource on which Arkansans placed high value—their wildlife. At the time, however, Arkansans were angry about the president's action. In 1909, in fact, the Arkansas General Assembly adopted a resolution in which it called on the state's congressional delegation to do all it could to see the two national forests "disestablished."

Early Big Lake National Wildlife Refuge sign

In 1915, Big Lake was designated as Arkansas's, and one of the country's, first national wildlife refuges.

Arkansas Game and Fish Commission.

PART 2

Politics and Exploitation

6 Early Wildlife Laws

In 1873, the Arkansas General Assembly noted the plight of the state's many fish-depleted streams first. The legislature enacted a law making it a misdemeanor for anyone to use poison to take fish from lakes, ponds, or waterways in Arkansas. The fine upon being convicted was from twenty-five to one hundred dollars, a stiff punishment for the times. Two years later, the General Assembly made it a misdemeanor for anyone to take fish from an enclosed or artificial lake or pond without the owner's consent, provided three signs had been posted warning the water was intended for breeding or cultivating fish. The fine was one hundred dollars.

Because market hunters and their activities were so visible, the state sought to curb their activities when, also in 1875, the legislature followed the lead of other states and imposed a tax on nonresident trappers, hunters, and seiners or netters of fish. At a special session only nine months later, the General Assembly amended this law to make the tax ten dollars and to require that nonresidents obtain licenses from county clerks in the counties in which they proposed to hunt, trap, seine, or net.

Concern about the state's generally declining fisheries was sufficiently great in 1875 that the house and senate adopted a joint resolution calling on the governor to appoint a three-man commission to study the possible introduction of sea fish, as well as new varieties of freshwater fish, into Arkansas rivers The commission also was to make suggestions about protecting fish after contacting fish commissions in other states doing restoration work. It was to report its recommendations to the 1877 General Assembly; however, no new laws affecting Arkansas wildlife were passed in that session.

In 1879, after outlawing the use of poison six years before, lawmakers took a bigger step. A law enacted that year made it illegal to drop or deposit "any explosive material or substance or any intoxicating or stupefying liquid, drug, vegetable or fruit" in any body of water for the purpose of killing, maiming, or paralyzing fish. The fine was from five to twenty-five dollars. The same law required persons owning or controlling dams or any other obstruction across a waterway to keep the structure "open sufficiently to permit of the free and easy passage of all fish either ascending or descending" rivers or streams between March 1 and June 1 annually. The law did not apply to dams built for water-powered mills and "manufactories." The legislature also amended a law so that anyone killing, maiming, or wounding an animal belonging to another would not be jailed or be liable for a twenty- to one hundred-dollar fine if the animal was trespassing on fenced property. The issue of trespassing by animals or people already was growing into a thorny thicket. Much of the livestock in Arkansas, particularly in the mountains, was not penned or fenced in the 1800s, though legislation enacted to define what constituted a fence indicates that more and more fences were being built in an effort to keep free-roaming animals out, not in.

It was another six years, in 1885, before the General Assembly's attention returned to the state's game and fish problems to any substantial degree. This time the focus was on deer, turkeys, prairie chickens, and quail, species already showing alarming declines. A closed season was established for prairie chickens. It was made unlawful to pursue deer with the intent to catch, kill, or injure them (or to chase with dogs with or without such intent) between February 1 and September 1. The closed season on wild turkeys was May 1 to September 1, and for quail, March 1 to October 1. The law did not apply to persons hunting or trapping turkeys, prairie chickens, or quail on their own land, especially if the birds were destroying crops. It also was made unlawful to destroy, disturb, or rob the nest of any wild bird except those of crows, blackbirds, hawks, owls, and eagles. During closed seasons, railroad companies, and steamboats were prohibited from possessing or receiving protected species with the intent of transporting them. Fines varied from three to ten dollars for each bird and destroyed nest and from ten to twenty dollars for each deer. Because the state had no game wardens, the legislature declared it was the duty of "all sheriffs, constables, coroners, marshals, market-masters and police officers" to enforce the law.

The people of Arkansas were in a dilemma. They could see game and fish disappearing, yet they were emotionally committed to the belief that government had no business tampering with their "God-given rights" to hunt and fish as they saw fit. That dilemma was reflected in this editorial published in 1885 in the *Southern Standard* newspaper in Arkadelphia:

> By the provision of the game law approved by the governor on the 23rd of February . . . it will be seen that section 4 of the act makes it unlawful for any persons whomsoever to have in their possession during the season inhibited by law, any of the game mentioned in the act, without regard to when it was killed or caught!
>
> Section 5 prohibits the transportation thereof during the months named in the act whether killed or caught previous to the time inhibited or not. Evidently this bill was not properly considered, or these features would never have been incorporated in it. We do not find fault with the main objects of the act, for we think it a most important and a very necessary provision of law, if it had been properly worded. What we do stand up on our "dew claws" and howl about as we think our citizens will when they understand the points of our objection is this: On the last day of January, the day before it becomes unlawful to kill or catch deer, hunters all over the state will kill from three to a dozen deer, as they may legally do, but under the provisions of the bill if he or his neighbors or the butcher or anybody else should be caught in possession of any of the game so killed the next day or at any period thereafter between the 1st day of February and the 1st day of September he commits a misdemeanor.

The dates of the closed seasons were adjusted in 1889, but nothing was done about the problem the editorialist perceived.

Few Arkansans were upset when the General Assembly enacted laws affecting the species they believed were destroying their livestock. Rather than dictating bounty amounts, the legislature amended the law in 1885 so each county determined how much would be paid to those who killed a wolf, wildcat, or panther and exhibited its scalp, "with both ears of said animal," to a magistrate within thirty days. Persons caught stealing a wolf, wildcat, or panther from a trap in order to deprive the owner of a bounty would have to forfeit five dollars for each wildcat, eight dollars for each wolf, and ten dollars for each panther, plus court costs.

By 1885, the legislature also was more serious about cracking down on what was seen as the cause of the state's fisheries decline—market fishermen. The law enacted in the 1885 session outlawed the use of any seine, net, gill net, trammel net, set net, bag weir, bush drag, and any fish trap, dam, or device designed to take fish in all waters of the state. The law did not apply to persons fishing in waters wholly on their property, and there were other exceptions: Seines not exceeding fifteen feet could be used only to catch minnows for bait or to stock other waters; a seine not longer than sixty feet could be used in any unnavigable stream to catch fish "for family use, or for picnics, or not for sale," and a trap could be placed in an unnavigable stream so long as it did not block passage of fish. Seines, nets, or gill nets with meshes of three inches or larger were not outlawed. Violations were misdemeanors carrying a penalty of five dollars to two hundred dollars. The law was tightened in 1889 to eliminate the exception of nets with meshes of at least three inches.

The 1879 law requiring dam owners to leave passageways for fish was being ignored, so the 1885 legislature became quite specific about what it wanted. The new law specified every dam was to have an apron or chute "not less than 14 feet wide and sloping from each side to the center, so that the center shall be at least 6 inches lower than either edge and having an inclination of not more than 45 degrees and so situated that the main current of water impeded in its natural flow by the dam or other obstruction shall pass over the same, or otherwise build a dam or obstruction with an apron or chute at its lowest point for the free passage of fish whenever the stream is swollen beyond its ordinary size." Each day that a dam did not have these features was to be a "distinct offense"—a misdemeanor with a fine of five dollars to twenty dollars. The law encountered stiff public opposition and was repealed two years later.

The 1885 session was also noteworthy because it brought passage of what eventually became a swarm of laws protecting various species of game and fish in individual counties. The honor of being first went to Yell County. Designed to preserve the county's game, the law banned killing, trapping, or netting partridges (quail) within Yell County's boundaries from March 15 to September 15 and wild turkeys from April 15 to August 15. Chasing, hunting, or killing deer was illegal from February 1 to August 1. Further, possession of these animals during closed seasons was prima facie evidence of guilt, with a fine of ten to twenty-five dollars.

Nothing the General Assembly did seemed to correct the fisheries decline. So, in 1889, legislators took the big step, already tried by some other states, of prohibiting export of game and fish from Arkansas for six years (extended in 1895 for six more years to 1901). The measure was aimed at stopping market hunters and fishermen. Section 1 of the law set forth a concept necessary to enactment of the law but also foreign to many Arkansans and unacceptable to some. It read

> That all the game and fish, except fish in private ponds, found in the limits of this State, be and the same is hereby declared to be the property of the State, and the hunting, killing, and catching of same is declared to be a privilege.

The 1889 law referred to game in general. In 1893, the law was amended to allow the export of beavers, opossums, rabbits, groundhogs, raccoons, squirrels, snipes or plover, ducks, and geese "when shipped openly." Two years later, the legislature reversed itself and put all these animals except ducks and geese on the list of species illegal to export. Also in 1893, the legislature said the ban on export was not to be applied to catfish or buffalo fish caught in Mississippi, Crittenden, St. Francis, Poinsett, Jackson, Phillips, Desha, Chicot, Lincoln, Prairie, Craighead, Greene, Clay, Ashley, Cross, Marion, or Jefferson Counties. The 1895 amendment expanded the exemption to include catfish and buffalo fish from Lafayette, Hempstead, Miller, Little River, Independence, Columbia, Faulkner, Calhoun, Pope, Woodruff, Randolph, Ouachita, Lawrence, Lee, Drew, Union, Sevier, Boone, Lonoke, Perry, Madison, Newton, Searcy, Fulton, Washington, Sebastian, Montgomery, Carroll, Howard, and Yell Counties. In 1895, specific waterways were first placed off-limits to seining, netting, and trapping fish. Leading the list was Big Spadra Creek in Johnson County.

To their dismay, legislators learned in 1897 that not even songbirds were spared from exploitation. They responded by amending a law so it was unlawful to kill, wound or injure any wild bird "other than the game birds," and to destroy, disturb, or rob their nests. The list of undesirable species to which this did not apply was expanded to include the English (house) sparrow as well as blackbirds, crows, eagles, and others named previously. Chinese (ring-necked) pheasants joined the list of protected bird species (for five years) in 1899, because they had been stocked in a few areas. The stockings proved unsuccessful.

Meanwhile, deer were becoming scarce in the Delta as levee building, followed by land clearing and conversion, gained momentum. In hopes of arresting the

White pelican

It was not until 1897 that nongame birds like the white pelican were spared from exploitation. That year Arkansas legislators passed a law making it illegal to kill, wound, or injure any wild bird "other than the game birds," and to destroy, disturb, or rob their nests.

Arkansas History Commission.

decline, the 1897 General Assembly enacted a law prohibiting chasing deer with dogs in Mississippi, Crittenden, Cross, Poinsett, and St. Francis Counties. The same law banned the discharge of firearms on lakes and streams in those counties after sundown. (This was amended in 1899 to say it was illegal to discharge firearms after sundown with the intent of killing, injuring or frightening, "any aquatic fowls.") Each offense carried a twenty-five- to fifty-dollar fine.

Mississippi and Crittenden Counties followed in 1901 with laws declaring it illegal to shoot deer, turkeys, or wild fowl on any posted property, open or enclosed, without the owner's written permission. In another 1901 law, the ban against chasing deer with dogs was dropped by Cross, Poinsett, and St. Francis Counties but retained by Mississippi and Crittenden Counties. But the same legislature made it illegal to chase deer with dogs in all of White County and in Barren and Glaize Townships of Jackson County. This law was repealed two years later. The use of dogs to catch, kill, or injure deer, or to pursue them without those intents, was banned statewide from March 1 to September 1 in a law enacted on May 23, 1901.

It was not unusual for the legislature to enact a law in one session and repeal it two years later because of adamant constituent opposition. This occurred in Conway, Perry, and Pope Counties between 1897 and 1899 with respect to a law that prohibited fishing with anything but a conventional hook and line everywhere in those counties except on the Arkansas River. As the General Assembly became even more involved in minute amendments to game and fish legislation, a law might be passed and repealed during the same legislative session. For example, the 1903 legislature took steps to regulate mussel and pearl fishing by passing a bill on March 9. Less than two months later, on May 20, the law was repealed.

In other ways, 1903 was a notable year for Arkansas wildlife legislation. One measure the General Assembly enacted made it illegal to catch or kill wild turkeys any time, anywhere in the state using blinds or baiting or by means of pens, trap doors, trenches, or deadfalls. Gov. Jeff Davis let the bill become law without his signature. The significant legislation enacted in this session, however, wrapped most of the state's major wildlife laws into one package and declared that the sheriff of each county (except Mississippi) or any constable was an ex-officio game warden, as were all of their deputies. Further, "it shall be the duty of the game warden or any constable or deputy, to make arrests, and prosecute offenders." If game wardens or justices of the peace received information about violations and failed to prosecute, then they were guilty of the misdemeanor of nonfeasance.

The law, Act 162, made the following activities illegal: trapping, netting, shooting, or killing quail or partridges from February 15 to November 1; shooting, killing or pursuing by dogs any deer or fawn from February 1 to September 1; trapping, netting, shooting, or killing any wild turkey from May 1 to September 1; shooting, hunting, fishing, or trapping by nonresidents; shooting after sunset or before sunrise on any waters of the state; shooting wild ducks and geese before sunrise or after sunset; and shipping, exporting, or carrying from the state any game fish or game of any kind. All conflicting laws then on the books were repealed.

Passage of statewide laws, however, just set the stage for individual counties to seek special legislation tailored to their local conditions. For example, Ouachita and Union Counties got a bill passed in 1905 to ban killing squirrels within their borders for the purpose of selling, bartering, or exchanging them for other goods. Under another 1905 law, nonresidents were allowed to fish (but only with hook and line) in the Spring River in Sharp and Fulton Counties.

The 1905 General Assembly also "adjusted" Act 162 of 1903 to exclude rabbits from the ban on shipping game out of state, provided the rabbits were shipped in bunches or in open crates that were "at all times open to inspection." Arkansans could not make up their minds whether rabbits were game; in addition, there was a big market for the animals in several eastern states. In another 1905 law, bears, squirrels, raccoons, and opossums were excluded from the ban on possession for sale. Monroe, Lee, Phillips, and St. Francis Counties were dissatisfied with this and obtained a law in 1907 that prohibited killing squirrels for any reason within their borders from December 1 to May 1 annually.

Also in 1907, the General Assembly authorized a comprehensive law for Desha County. The law, Act 269, provided that it was unlawful to hunt or trap in Desha County on another's property without paying a fee and obtaining a license from the county clerk; licensed persons could invite nonresidents to hunt; hunting and trapping were restricted to the seasons set in state law; and there was a bag limit on deer. In effect, the Desha County law was the forerunner of Arkansas's first state Game and Fish Commission and the laws under which it operated, which included statutes on licensing and bag limits. Phillips, Chicot, Bradley, Dallas, and Stone Counties followed with similar laws in 1909. Licensure to fish in the Western District of Clay County was authorized in another 1909 law.

The 1909 General Assembly enacted at least thirteen laws relating to wildlife, virtually all of them dealing with individual counties or clusters of counties that either wanted more stringent protection of their game and fish or liberalized exemptions from state laws. A moratorium on such time-consuming changes was in effect during the 1911 General Assembly, which was in the process of moving into the new State Capitol from the Old State House, but the floodgates opened again in 1913. The demands for exemptions for this or that county or this or that body of water became so great that the 1915 General Assembly enacted legislation to create the state's first Game and Fish Commission. This was not an easy decision for the General Assembly, even though the shortage of game and fish obviously was growing more acute each year. The concept of government regulating hunting, trapping, and fishing was a hotly debated topic throughout the period.

Tom Wallace, chief of the *Louisville Times* editorial staff, told a story in his speech before the National Game Conference in the early 1900s (published in the *Arkansas Conservationist* October/November/December 1924) that was just as applicable to Arkansas as to Kentucky. Wallace confessed that the only bill for which he ever personally lobbied the Kentucky legislature was one for game protection. He recalled,

> Sitting as the representative of one of the western counties, was a man with billygoat whiskers who in private life divided time between the pulpit and the plow. So infuriated was he at the idea that anyone should consider abridgement of his liberty that he said he would shoot quail on the nest if he cared to do so on his own land, and that—with a good round oath—he would shoot hell out of the first son of a delinquent who should attempt to interfere.
>
> "My dear sir," I protested. "You are a minister of the Gospel and a lawmaker, yet in language shockingly profane, you express sentiments wholly lawless."

> But he would not be rebuked. He desired to fight. He made forthwith a proposition looking to immediate action. What followed has no bearing upon this discussion.

Wallace told the gathering that it was in this spirit that early efforts to pass game laws were met almost everywhere in America, and the spirit survived in some sections. "The exaggerated individualism, the grotesque egotism of the ruralist, which time and education will wear down, has stood greatly in the way of game protection," Wallace said.

Sharon Bass said it well in *For The Trees,* her work on the history of the Ozark and St. Francis National Forests:

> Hunting and fishing to the Ozark mountaineer, or would-be mountaineer, is more than sport. It is survival and it is ritual. Hunting might rate as the 11th article in an Ozark bill of rights. Hunting in Arkansas is perceived as a God-given right, a joy, a perennial call to the wild, with which neither man nor government dare interfere . . .

The Ozark sentiments easily could describe the spirit that prevailed throughout the entire state.

7 Beginnings of the Conservation Movement

By 1900, the United States was one of the greater economic powers in the world. Its population stood at more than 75 million. With all this growth, however, came the slow realization that the country's strength—economic and cultural—was based on its natural resources, including wildlife: "Monuments should be erected to the deer, bison, elk, antelope, sheep, quail, turkey, prairie chickens, and wild ducks and geese, to commemorate their services as nation-builders and wealth-producers," Dr. William R. Hornaday, director of the New York Zoological Park, declared in 1923. "If their value up to date could be figured out, think you that it would fall one penny short of $100,000,000?"

The nation was steeped in materialism and exploitation. An appreciation of scenic beauty, or even a sense of stewardship of the land, was not part of the spiritual makeup of America's settlers. Most seemed to agree with early-American clergyman Cotton Mather's pugnacious contention that "What is not useful is vicious." Mather based his view on Genesis, which commanded man to "have dominion over the fish of the sea, and over the fowl of the air, and over the cattle, and over all the earth . . . and replenish the earth and subdue it."

Early Americans added their own unique political heritage to this Biblical commandment. Men had been restrained in the Old World. Much of the land, and the wild creatures on it, belonged to the nobility and was forbidden to the majority of people. Poaching was the only way common men could tap those natural resources. Once those fetters were thrown off, men took what they could without fear of authority or concern for tomorrow. If they gave any thought at all to the situation, most Americans assumed it would never end—that natural resources were infinite and would last forever. Further, this idea appears to be a constant in the American character.

For example, in 1973, the *Stuttgart Daily Leader* in an article titled "Early Hunting Practices Recalled" interviewed a man who came to Arkansas's Grand Prairie in the early 1900s. The subject was hunting practices in those days. As the man, unidentified at his own request, reflected on the large duck kills in those early days in the islands of woods dotting the prairie, he was "quick to point out that it was a situation much like the energy crisis today: There were so many millions of ducks here

Large flock of ducks near Stuttgart

In the early 1900s, ducks wintering on the Grand Prairie near Stuttgart gathered in flocks numbering in the millions. Despite the fact that Arkansas's vast flocks of passenger pigeons had been hunted out years earlier, most people who lived at the turn of the century still considered the supply of game in Arkansas inexhaustible.

Arkansas Game and Fish Commission.

that they believed that there would always be a plentiful supply of ducks, geese, and prairie chickens; just as up until recent months the average person thought there would continue to be an abundance of oil since there always had been."

Then as now, a few persons thought differently. Some of these, like Gifford Pinchot, had attained positions in the federal government by the late 1800s, which enabled them to effect changes and popularize their beliefs. Some conservation historians believe the beginning of change sprang from a dedicated bureaucracy. Others challenge this, saying it was based on a "narrow concept of what conservation is," and that private individuals working through organizations of their own were the movement's real originators (Graham 1971).

Webster defines conservation as "planned management of a natural resource to prevent exploitation, destruction, or neglect." Those conservationists who in the 1880s established the agencies that eventually grew into the U.S. Fish and Wildlife Service and the United States Forest Service are at least credited with giving the movement a name that helped make it a revolution—a revolution that has progressed in three distinct waves in the United States during the last one hundred years, according to *Highlights in the History of Forest Conservation*.

Pinchot, the father of modern forestry in America, knew the movement needed a name if it was going to become popular and gain broad-based public support. It was Pinchot and his fellow scientists whose influence was so pervasive and whose ideas were so successful that even a modern historian can write, "It is from the vantage point of applied science, rather than of democratic protest, that one must understand the historic role of the conservation movement" (Hays 1959). These scientists met regularly to talk about their ideas, including management of the newly created forest reserves and the need for naming the movement. During one discussion, someone mentioned the government-owned forests in India were called "conservancies." Pinchot's assistant, Overton Price, suggested the movement should be called "conservation," and President Theodore Roosevelt agreed. "So we called it the conservation movement," Pinchot said (Graham 1971). No matter who coined the phrase, it was Pinchot and Roosevelt, especially the latter, who probably contributed more than any others to make this first wave of conservation important in the country's history.

Despite Pinchot's imposing presence in much of the conservation activity during Roosevelt's administration, there were certain areas in which the forester showed little interest or took no part. Officially, wildlife did not concern him. But whatever initiative Pinchot lacked was more than made up for by the passionate interest of the president himself. As Dr. Hornaday, the New York zoologist, later commented, Roosevelt "gave the vanishing birds and animals the benefit of every doubt." With the destruction of a species, Roosevelt felt "as if all the works of some great writer had perished," and the destruction of the country's native fauna was at its peak when he entered the White House (Graham 1971).

Theodore Roosevelt

Destruction of America's wildlife was at its peak when Theodore Roosevelt became president. The first significant wave of the conservation movement began during his administration, which oversaw the creation of fifty-one national wildlife refuges and eighteen national parks.

Arkansas History Commission.

During his administration, Roosevelt enlarged the size of the national forests from 42 million to 172 million acres, adding the Ouachita and Ozark National Forests in Arkansas, in 1907 and 1908. He created fifty-one national wildlife refuges and set aside eighteen areas of special interest as national parks, including the Grand Canyon and the Petrified Forest. Much of this land was set aside in the administration's closing weeks, as Roosevelt and Pinchot worked frantically to nail down their conservation gains because they suspected the first wave was playing out, and the president's successor, William Howard Taft, would not go forward with the conservation movement. They were right. Another significant event took place in the waning years of the Roosevelt administration—the holding of a White House Conference on Conservation that spread the new word "conservation" and its concepts. The conference inspired forty-one states to form conservation commissions of their own.

The second wave of conservation occurred during Franklin D. Roosevelt's administration. Chief among the conservation efforts at the time was the work of the Civilian Conservation Corps (CCC). Between 1933 and 1942, more than two million men helped to protect a vast amount of forest: planting trees, restoring watersheds, controlling erosion, developing recreation areas, and making other improvements. The third wave began about 1960 and extended through the 1970s.

Compared to the eastern states, Arkansas was young when the first conservation wave rolled across the country, and timber companies were doing large-scale leveling of forests in the Delta, the Coastal Plain, the Ozarks, and the Ouachitas. Improvement districts were forming, and new levees and drainage canals were constructed almost daily in the Delta. Market hunters already had been there, leading to the array of laws against the export of Arkansas's game.

The clamor against the "unremitting butchery" by "sportsmen" and market hunters arose throughout the country, and nowhere was it more insistent than among ethical sportsmen themselves. George Bird Grinnell's *Forest and Stream* was in the forefront of the reform movement. More spectacular, if not quite so effective, was G. O. Shields, the outspoken editor of *Recreation Magazine*, who coined the phrase "game-hog." The market hunters did not play dead; there were always bitter legislative battles over game-protection bills (Graham 1971).

The combination of overhunting and habitat alteration and destruction made for a particularly critical wildlife situation in Arkansas just as the first conservation wave was cresting and receding. The situation was not reversed until the 1950s.

It was a situation that caused Gurtha Taylor to cry out in a prize-winning essay at the Ashley County School Fair in 1925, published in the *Arkansas Conservationist* in 1925:

> Human hands never fabricated a more finely balanced bit of machinery than nature's provision for perpetuating wildlife.
>
> The state of game hunting in North America is a maze of contradictions . . . American sportsmen are wide awake and sound asleep. The laws to protect never before were so many nor so good, and yet game was never more certain than now of extermination on a wide scale.
>
> . . . our fears for the future will not be kept silent . . . We view with dismay the tremendous power of the destructive influences that are now in operation not only to destroy the game of North America, but to sweep all legitimate hunting grounds absolutely bare of fur, fin and feathers, hoof and horn.
>
> There are five million guns actually active in the limited states against game. The deadliness of modern sporting firearms now is amazing and appalling.
>
> The game of the world has less cover, less food, and less breeding security than ever before. The really savage element in the army of destruction is just as brutal, savage and relentless as ever it was in the worst days of the past. The real game hog is quite as relentless as ever.
>
> . . . the very democracy under which we live and to which we point with pride actually makes against any such practical handling of the problem. Today each man owns his land in fee, and he is lord thereon. The owner cuts his forest with no thought of the results to the state, and the denuded hills precipitate the waters to the sea. As the streams dry up and the ponds and lakes are drawn on to turn the wheels of commerce, their receding waters leave the valuable food-bearing vegetation on the shores to wither and die.

The "appalling" weapons to which Taylor referred were rifles and double-barreled, breech-loading shotguns that had replaced muzzleloaders in the hands of hunters of the late 1800s. Despite Arkansas's rapidly declining game supply and the voicing of such sentiments as Taylor's, the state was not caught in the crest of the conservation wave.

A group of state and federal wildlife administrators met with their Canadian counterparts in 1902 at Yellowstone National Park and formed the International Association of Game, Fish and Conservation Commissioners. The organization soon became a strong force in the conservation movement, doing much "to raise professional standards within its own sphere of influence" and fighting for federal legislation. The years between 1900 and 1915 saw more substantial progress in the field of wildlife administration than had taken place in the preceding century. By 1911, the number of states without central administrative agencies had shrunk to seven—Arkansas, Florida, Mississippi, Nevada, Virginia, North Carolina, and Washington. Most of these had administrative programs by 1915. The backwash of the initial conservation wave reached Arkansas in that year with the creation of the state's first Game and Fish Commission (Trefethen 1961).

8 Early Years of the Game and Fish Commission: 1915–1944

STATE DEVELOPMENTS

Gov. George Washington Hays signed the bill that became Act 124 on March 11, 1915. The *Arkansas Gazette* called it the "first real game law Arkansas has had," presumably because "no county is exempt." J. M. Futrell of Paragould was given credit for much of the leadership that resulted in Act 124: "among the leaders of a movement to set up a conservation agency shortly after 1910 was J. M. Futrell, state senator from Paragould who later became governor." The *Arkansas Gazette* reported that the main features of Act 124 were proposed in a bill by Senator Futrell, but the measure that finally passed was from the senate judiciary committee. The bill, the newspaper said, "had a strenuous career," meaning it passed only with great difficulty.

The official magazine of the Arkansas Game and Fish Commission said in a 1974 article that

> Not long after its creation [in 1915], the commission lauded Futrell with this resolution: "To the excellent services of Governor J. M. Futrell, a member of the Senate of 1915, we owe the credit of the passage of the act creating the Game and Fish Commission. While there were many good members of both the House and Senate who took an active part in the passage of the act, Governor Futrell took the initiative."

Futrell did not become governor until 1933.

The emphasis in the act was where it had been in laws passed by previous legislatures—on setting seasons and bag limits for certain species. It defined the open season for deer, bears, and turkeys as the last twenty days of November through the first ten days of January. Turkey gobblers also could be hunted legally from April 15 to May 31 each year. The license for deer was one dollar, and it was illegal to hunt

Gov. Junius Marion Futrell

During his term as state senator, J. M. Futrell of Paragould led the effort to create a state conservation agency. His work led to the passage of Act 124, which established the first Arkansas Game and Fish Commission on March 11, 1915. Futrell became governor in 1933.

Arkansas History Commission.

deer without a license. The bag limit for a season was one bear, two deer, and four turkeys. Trappers using more than ten traps were required to pay a ten-dollar license fee. Beavers and otters were protected for five years. December and January were the open season on quail, with a bag limit of fifteen a day. The bag limit on ducks was twenty-five a day.

It was a violation to use a dog to hunt birds if the dog had not been licensed for one dollar. Further, it was a felony to steal a licensed dog, regardless of whether or not it was a bird dog. Nonresidents could not be licensed, nor anyone convicted of violating the law twice. No person convicted of larceny or a habitual drunkard could be licensed to hunt or fish under Act 124. The law prohibited hunting on enclosed premises without the owner's written permission.

Market fishermen were prohibited from plying their trade unless they obtained a twenty-five-dollar license, and no one was permitted to use a seine or net with mesh smaller than three inches square. Persons using seines for family or picnic purposes were permitted to use one not exceeding sixty feet in length with mesh not smaller than three inches square. Artificial bait users were required to pay a one-dollar license fee. It was a violation of the law to ship or export game or fish out of the state.

Act 124 made it the duty of the prosecuting attorney to file information against persons violating the law; upon conviction, he received a twenty-five-dollar fee. If the case was appealed to the circuit court, the prosecuting attorney received an additional ten-dollar fee. Prosecuting attorneys were directed to make semiannual reports to the Game and Fish Commission.

Creation of the commission itself was treated almost incidentally. For example, the *Arkansas Gazette* said in the third paragraph of its report on the new law that its principal features included "a system of game wardens and deputies, supported by the licenses and penalties, a bag limit for quail and ducks, shortened hunting seasons and a commission to serve without pay." The commission was to consist of five members appointed by the governor for four-year terms. Two of the first five members, however, were to serve only two-year terms. "Only such persons as are interested in and devoted to the preservation of wildlife shall be appointed to serve,"

Early Arkansas Game and Fish commissioners

The original Arkansas Game and Fish Commission and game wardens. *Seated left to right,* Commissioners Dr. Horatio Wells, Monticello; George Webber, Texarkana; C. C. Calvert, Ft. Smith; Judge Lee Miles, secretary, Little Rock; and D. G. Beauchamp, chairman, Paragould. *Standing left to right,* game wardens, Foster Rogers, Texarkana; Med Donaldson, Lake City; Joe Galvin, Stuttgart; Herschel Neely, Paragould; Clyde Marsh, Prescott; Wash Clibourn, Conway; James Fernandez, Fort Smith; George Rison Jr., Little Rock; and Ed Huddleston, McGehee.

Arkansas Game and Fish Commission.

the law specified. When and if the commission collected enough money from fees, the members' expenses traveling to and from meetings could be paid; otherwise, they were to receive no compensation. The commission was

> To promote and assist in stocking private and public waters with fish, encourage private and public game preserves, but are not authorized to expend any funds upon this account.
>
> To supervise and direct the enforcement of all laws relating to the protection of fish and game, their authority to be concurrent with, and not exclusive of, the authority now vested by law in other officers.
>
> Vested with such power and discretion in making rules and regulations as may be necessary for the proper enforcement of the Act. [However, the legislature retained the authority to set seasons and bag limits.]
>
> Directed to appoint a secretary for the commission and fix the pay for his services, prescribe his duties, and could delegate such powers to the secretary as it deemed necessary, not inconsistent with the provisions of the law.
>
> Told it could appoint such persons as may be necessary for the proper enforcement of the laws, and shall have power to discharge any of those persons with or without cause. These persons were to be known as game wardens, and those appointed had to be at least 21 years old and be of "good moral character." A person appointed as a game warden also had to post a $1,000 bond "conditioned for the faithful performance of his duties."

At the commission's first meeting, D. G. Beauchamp of Paragould was elected chairman, and Judge Lee Miles was chosen as secretary. The other commissioners were Dr. Horatio Wells of Monticello, George Webber of Texarkana, and C. C. Calvert of Fort Smith.

As soon as the commission was organized as a division of the state government, it turned its attention to arranging a warden system. "After conferring with the game and fish departments of a number of other states," according to its first annual report, the agency reached the opinion "that the best service could be had by a warden who could devote his entire time to the work. We therefore decided to select as many wardens as the fund accumulated would justify, and instruct them to patrol the state under the direction of the secretary." But the commission had no previous experience judging how much might be accumulated in the fund and, by its own admission, was "conservative" in its approach.

Initially, the commission went on to explain, it thought it could employ a man, on a small salary, in each county and "have him devote a portion of his time to this work, but this proved to be unsatisfactory in cases where we tried it." In addition to describing the warden system it put in place, the commission boasted in its first annual report that, "We now have a good many volunteer sportsmen and conservationists from every section of the State who report to us any violations . . . and other things needing our attention."

Part-time wardens hired by the commission were Joe Galvin of Stuttgart, James Fernandez of Fort Smith, Med Donaldson of Lake City, Foster Rogers of Texarkana, Wash Clibourn of Conway, Herschel Neely of Paragould, George Rison Jr. of Little Rock, Clyde Marsh of Prescott, and Ed Huddleston of McGehee. The commission said it also "had some special men employed at various times in cases of emergency, such as overflows along the Arkansas, Cache, White, St. Francis, Mississippi and other rivers."

In its synopsis of Act 124, the commission said only that its wardens or their deputies had the right of law "to search any package, express, freight or otherwise, for violations." The commission did not note that a key shortcoming of the law was that wardens (through 1930) had no power of arrest and were poorly funded. In order to arrest a violator, a warden had to be accompanied by a peace officer who made the arrest. In the absence of a peace officer, a violator could be apprehended only if the landowner, on whose property the offense had occurred, would swear out a warrant. In court, only the prosecuting attorney could prosecute the case.

The unidentified author of an article in the 1974 commission magazine declared it was hard to understand why a man would have wanted to serve as a game warden during the 1915 to 1930 period. A warden's salary in 1915 was about eighty dollars a month, and transportation was not provided. If he used a horse, he was allowed only a few dollars for feed. Being a game warden, even one who was not working for the commission, was dangerous, too. The *Arkansas Gazette* reported on March 1, 1915, that Walter Gleabes, forty-five, a deputy U.S. marshal and warden for the Big Lake Shooting Club, had been shot from ambush for the twelfth time. A bullet from a .35-caliber gun passed through Gleabes's left lung the night of February 28, and he was not expected to live. The shooting, the newspaper explained, resulted from a feud between residents of the area and the owners of the club, who were not Arkansas residents. "All of the shootings, it is said, have been the result of attacks upon Gleabes because he was employed by the club and has made arrests of trespassers for violations of the club's hunting rules," the article said.

Game warden J. W. "Bill" Smith

Smith is pictured with confiscated fishing tackle seized in a lake near Dyer in 1926. Game wardens of this period were handicapped because they did not have the power of arrest. Arrests were made by peace officers accompanying the warden.

Arkansas Game and Fish Commission.

Under Act 124, all fines assessed by the courts were to go, after certain deductions, into the commission's general operating fund, known as the Game Protection Fund. After the prosecuting attorney received his fee in a conviction, a large slice of the remaining fine went into the local school fund. "Thus, only a slim amount of the original fine actually went into the Game Protection Fund. This, coupled with the general lack of revenue, made the way even more difficult for the fledgling Commission."

In 1915, the commission received $6,350 for commercial fishing licenses, $4,808 for artificial bait licenses, $3,850 for deer hunting licenses, $2,350 for dog license fees and $320 for trappers' licenses. The Game Protection Fund realized a meager $163.34 from fines. Only $8,460.82 of the total $17,841.34 collected was available for the commission to spend because the appropriation legislation was insufficient. Collections in 1916 amounted to $22,677.15. With a balance of $9,380.52 from 1915, the commission had a total of $32,057.67 available to spend. However, a taxpayers' suit was filed in Pulaski County, challenging the commission's legality. The state auditor was enjoined from issuing any warrants to pay commission expenses until a decision was handed down in the agency's favor on January 29, 1916.

Fish trap basket

Fish trap being destroyed

Unregulated commercial fishing caused drastic declines in the fisheries of many Arkansas streams by the 1920s. The photo *above* shows Warden A. E. Denton with the basket from an illegal fish trap confiscated on the Saline River in the 1930s. *Left,* Warden Victor Wyer destroys a fish trap near Danville in May 1932.

Arkansas Game and Fish Commission.

The commission's income rose slowly during the 1920s until, on June 30, 1930, it broke the $200,000 barrier with $205,003.48. "I know it seems little today, but then it was a memorable occasion," Nell Taylor told Judge Lee Miles in her report, *Report on Income and Disbursements,* prepared in 1957.

The new commission felt Act 124 fell short of the state's needs and prepared a bill for consideration by the 1917 General Assembly. Among other things, the commission recommended establishing a five-year closed season on prairie chickens; protecting squirrels from March 1 to September 1 or October 1 and setting a bag limit; prohibiting the sale of squirrels just as the existing law barred sale of ducks and quail; lowering the bag limit on ducks from twenty-five to fifteen and putting a bag limit on geese; banning the killing of all antlerless deer; and allowing the commission to establish hatcheries for fish and game birds.

In addition, the commission said that the sale of fish, game, and game birds raised in captivity on private premises should be permitted; that it should be a felony to use dynamite to kill fish; that it should be unlawful to kill, catch, or capture a furbearing animal at any time except between December 1 and February 15; that there should be a general gun license; and that the commission should be able to grant permits to game and fish departments in other states "to make scientific investigations for experimental purposes, and to permit the shipping of a limited amount of game and fish under the supervision of the Game and Fish Commission for stocking purposes."

The commission also pointed out to the legislature that Arkansas was the only state that refused to license nonresidents. It felt Arkansas should not be the exception. Hunting licenses for nonresidents, the commission said, should be at least fifteen dollars, and the fee for out-of-state fishermen (hook and line only allowed) should be not less than five dollars.

The commission was so pleased with what the 1917 General Assembly did, it proclaimed in its second annual report that the new law, Act 133, "very complete" and, with some slight amendments the agency intended to recommend, it would be "as good a law as that of any state in the Union." If the law was "very complete" as the commission claimed, it probably was because the members did not realize that a few of the changes granted in 1917 merely reflected laws that had been in the Arkansas code at one time or another from 1885 to 1915, and that the 1917 legislation amounted to regaining lost ground. For example, the use of all seines, trammel nets, and gill nets was prohibited again, and it was made unlawful to shoot over waters of the state before sunrise and after sunset. On the other hand, the nonresident licenses the commission recommended were granted, and a compromise resulted in bag limits being dropped from twenty-five to twenty on ducks and quail rather than the fifteen suggested. The commission did not get its December 1 date on furbearers and settled for October 1.

The 1917 law contained many new features. Dealers had to keep records of the fish they bought and pay one-sixteenth of a cent per pound to the Game Protection Fund. A creel limit was put on game fish. The hunting license was extended to cover bears, wild turkeys, and deer and to dogs used for hunting deer, bears, and turkeys, as well as game birds. Deer (except those with antlers of five inches or more), wild turkey hens and pheasants, grouse, prairie chickens, woodcocks, turtle (mourning) doves, and robins were protected until January 1922. Hunting deer, bears, turkeys, wild fowl, and game birds with a headlight or torch was prohibited. A closed season was authorized on fish during the spawning season, with the commission to set the dates. In its second annual report, the commission explained how it had approached this new responsibility:

> In 1918, in order to fix the spawning season as short a time as possible and at the same time protect the fish during the time that they should be protected, it was decided to place a number of the wardens on the rivers and waters of the state, and, with the cooperation and assistance of reliable and experienced fishermen, to make whatever investigation necessary to determine exactly when the season should be closed and when it should be opened.

The commission said this plan proved satisfactory, and some valuable information was obtained, including knowledge that buffalo, drum, and carp spawned between March 15 and May 5. However,

> The Commission realizing the food value of fish and the necessity of using our natural resources for substitutes for meat and other foods which necessarily must be sent to our armies and allies in Europe, decided to make the closed season as short as possible and the market fishermen were allowed to place their nets in the water and resume their fishing on April 5.

Also, in its second annual report, the commission continued its campaign to establish hatcheries and farms, saying, "practically every state" in the country had a game farm and several hatcheries. "It is our idea that we should have a farm and fishery combined," the agency asserted. "It is possible to get a good location" for a combined facility, it said, and added that "the toll taken by the hunter of game birds throughout the state has been such as to eliminate the breeding stock from certain sections" of Arkansas. "These could be propagated on a game farm and liberated for restocking in these territories." It also was in 1917 that the legislature, just two years after the commission's creation, reverted to passing of laws applying only to certain counties. Act 276 declared it to be "unlawful for any person or persons to sell or offer for sale any fish of any kind taken from the waters of Hot Spring and Garland counties in any manner that fish are now caught or hereafter may be caught." Lee County succeeded in 1921 in getting a law (Act 391) passed that set different seasons for several wildlife species within its borders. This was a prelude to disaster because, instead of amending one law—the one governing the Game and Fish Commission—the 1923 General Assembly enacted 32 separate laws relating to wildlife, ranging from one that permitted using seines and nets to catch game fish in "certain streams and bayous" of Chicot County to one that designated the time when quail could be killed in Columbia County.

The first major sign of the legislature's growing unhappiness with the agency form of government for wildlife surfaced in 1919, when it limited the number of wardens to eight. In 1921, the General Assembly enacted Act 209, which required license fees collected by many agencies to be deposited in the general revenue fund. In the Game and Fish Commission's case, the new law did not specify the money was to be placed in the Game Protection Fund within the general revenue fund as before.

The commission's fifth biennial report (1924–25), prepared by Secretary Guy Amsler, revealed that by 1924, the commission's patience had reached the breaking point. The entire report constituted what might be described as a scolding lecture aimed at the legislature. Some examples are

> At practically every session of our General Assembly since 1915, important changes have been made in the original Act. Unfortunately, many excellent features contained in the primary measure have been abandoned. Four repealed sections that have especially hampered the work are (1) Section 12, giving the Commission the right to employ the number of game wardens necessary to successfully carry on the work; (2) Section 13, vesting the Wardens with police power; *(3) Section 54, fixing a spawning season for all fish and (4) Section 55, placing a bag limit of 25 on all game fish. It can be said, without fear of successful contradiction, that wildlife conservation in Arkansas would be 10 years ahead of where it now is had these features remained on our statute books. It is nothing more or less than another case of personal activities refusing to recognize the laws of nature or respect the rights of coming generations.
>
> From time to time there have been rumors to the effect that the next legislature would put an end to conservation work in the State, yet several legislative sessions have convened since that rumor became rife and each time, able legislators from every section of the state have concluded that wildlife conservation was in keeping with our advancing civilization and necessary to the happiness and prosperity of our people. It has long since ceased to be a question of expediency and has become a more practical question of effective methods to be employed. Not shall, but HOW SHALL we save our game and fish, is the question arising from every nook and corner of the state.

---

*No factual basis for this could be found; Section 13 of Act 124 of 1915 sets the quail season.

> While we are able to show some progress in this report it is unfortunate that we are not in a position to show a great deal more. Limited means, ineffective laws, and lack of cooperation on the part of the general public and some local officials may be said to materially effect [*sic*] conservation work in Arkansas. Entirely too much explaining, persuading and educating is required before action can be had. If Arkansas' wildlife is to be saved and the "seed" we now have utilized in the most effective way possible, productive action rather than pacifying non-perpetuating methods must be resorted to. After 10 years of "word swapping," concerted action is in order. It is high time we were laying aside personal animosity and following precedents set by states that are successfully and intelligently conserving their forests, game and fish.
>
> At the last regular session when the legislature appropriated $80,000 per year for this department, there was some criticism of members who voted to spend this much on wildlife conservation, and since that time certain individuals, desiring to further their political or personal ambitions, have branded this as an "outrageous" expenditure of the taxpayers' money. Certainly any criticism in this respect is unjustly founded. In the first place, the men who hunt and fish pay for perpetuating field and stream; and in the second place, this department has never spent $80,000 in any one year—it has never collected that much during a year, consequently, didn't have it to spend.
>
> If the people want seining or netting stopped in certain mountain streams or lakes of the state, why have a whole raft of special acts when that duty could be placed on the Department by a very short general measure? . . . why enact a cut and dried special measure against netting and seining in those waters, thereby aiding the increase of gar, turtles and other game fish destroyers? A simple general act placing the responsibility on the department of saving the game fish and eliminating their enemies, would get far better results, be less expensive to the State at large, and relieve the individual Representative of a great deal of criticism by placing the Commission between him and his people and authorizing it to relieve any strained condition that might arise.
>
> The future of conservation work in this State is rather problematical. Just when an effective system will be resorted to is a little indefinite. History of wild life protection almost invariably reveals that effective work is actually begun when forests and streams are completely depleted. Arkansas is following the well trodden path. Being an inland state, and for a number of years sparsely settled, civilization, with all of its destructive features did not reach our wild life as early as toll was demanded in some of the older coastal States. Significant in this connection is the fact that States having a fairly heavy and general population are now operating under a workable wild life conservation system. Our "seed" is criminally close to gone and cannot possibly last much longer against an ever increasing army of hunters and fishermen.

Instead of angering the legislature as one might expect, the commission's report apparently made the lawmakers somewhat contrite. Only three special laws were passed in 1925, and by 1927, all wildlife-related laws were general in nature. In addition, the commission had obtained the authority to operate a state game farm, which was established in Benton County near Rogers.

An article in a 1979 issue of *Arkansas Game and Fish* magazine stated that the "only lasting accomplishment of the entire 30 years" by the previous commission

was its fish hatchery system. In its sixth biennial report in 1926, the commission said it acquired rearing ponds in various sections of the state to determine their value in producing native species of fish, particularly black bass and bream. By 1928, it had found that a "decided majority" of these ponds had proved "of little value for fish culture purposes" and the principal reason was that most were too small. No more sites were going to be leased or acquired, the commission said, unless they were large enough to "guarantee ponds of 10 to 30 acres."

Ponds were abandoned at Warren, Magnolia, Hot Springs, Conway, Mena, and Williford. However, waters at Springdale and Mountain Home had managed "very good production," and the commission thought there was "a good chance" of improving these from year to year. The only new pond completed during the biennium was a five-acre facility leased from M. L. Sigman at Monticello. Though small, the pond might prove productive, the commission said, if some repairs were made.

The "outstanding rearing pond project in the state" had just gotten underway near Centerton in Benton County. Benton County residents had furnished a thirty-four-acre tract on which a pond covering sixteen to twenty acres would be constructed. "One of northwest Arkansas' finest springs is to furnish ample water, while the surplus flow will be taken care of by a canal constructed around the east and south sides of the area. Concrete standpipes at the upper and lower ends of the pond will guarantee complete control of the water at all times," the commission stated in

Rearing pond at Warren

By 1926, the Game and Fish Commission was operating several rearing ponds around the state for black bass and bream propagation. Two years later, the commission reported that a "decided majority" of these ponds had proved to be "of little value for fish culture purposes."

Arkansas Game and Fish Commission.

Milk can fish stockings

Fish stocking was a prominent commission activity by the late 1920s. Cooperative efforts between the Game and Fish Commission and the Bureau of Fisheries led to stockings like those pictured. *Below,* Henry Korn and Evel Walthall stock fish on May 21, 1929, on the North Fork of Illinois Bayou in the Ozark National Forest. *Right,* President Barton of the Forrest City chapter of the Izaak Walton League stocks fish in Otter Pond Lake.

Arkansas Game and Fish Commission.

its seventh biennial report. All in all, however, two years of experience had "created serious doubt if rearing ponds located throughout the state have any particular value over numerous ponds constituting a fully supervised hatchery," the commission said.

Construction started in early June 1928 on what was billed as the "World's Largest Fish Hatchery." Located near Lonoke, the site comprised 266 acres, 250 of which were to be converted into ponds. Plans called for twenty-four ponds of one acre each with the remaining 226 acres to be made into ponds of 5 to 15 acres. Construction was expected to take two years, perhaps longer. The hatchery site had two wells, an electric motor, two dwellings, and other outbuildings and was expected to cost $18,000, with an additional $17,544 in materials to be paid for from the agency's biennial appropriation.

During late summer and fall of 1929, sixteen one-acre ponds were completed that produced 405,000 largemouth bass, 600 rock bass, 4,500 crappie, 103,000 bream, and 4,000 flathead catfish. By spring 1930, eight additional one-acre, eight two-acre, and four six-acre ponds and one seven-acre pond had been added, making a total of seventy-one acres for fish production during 1930. From the 1930 hatch, about 514,800 bass fry and a quantity of bream had to be stocked prematurely because of a lack of pond space in which to raise them to fingerling size. "It is hoped that the time is not far distant when all fish will be held until they reach a size which guarantees a high percent of survival when planted," the commission commented in its eighth biennial report. As the biennium closed, there were about 200,000 black bass and 500,000 bream at the Lonoke hatchery to be raised to advanced fingerling stages for fall and early winter distribution.

Lonoke hatchery construction sign and buildings

Some work designed to increase Arkansas's "rather depleted supply" of commercial fish had been undertaken, the commission reported. During the spring of 1928, the Bureau of Fisheries operated a buffalo fish hatchery at Benzal on the White River. The commission said the results were "entirely satisfactory" even though plans for hatching 50,000,000 eggs failed. A peculiar season with rapidly changing water conditions made getting eggs "extremely difficult" and only 8,392,000 buffalo were hatched and stocked.

The commission said it was going to try teaching commercial fishermen the proper method of taking and fertilizing buffalo eggs. "The plan is comparatively simple and can be done by the average fisherman right in his boat when ripe fish are caught," the report asserted, adding that a man would be sent to "instruct fishermen desiring to cooperate" in the spring of 1929.

Construction on the Game and Fish Commission's Lonoke hatchery, billed as the "World's Largest Fish Hatchery," began in 1928. The hatchery consisted of 266 acres, of which 250 acres were to be made into ponds. The first crop of fish was ready in 1929. The hatchery was later named the Joe Hogan State Fish Hatchery after the first supervisor of fisheries.

Arkansas Game and Fish Commission.

Aerial view of Lonoke hatchery ponds and willows

Buffalo fish hatchery operated by the Bureau of Fisheries on the White River at Benzal, 1928

Arkansas Game and Fish Commission.

Arkansas received and stocked three shipments of rainbow trout from the Bureau of Fisheries. Two carloads were placed in Mammoth Spring and Spring River; the third went into Little Buffalo River in Newton County, the commission said. Men with trout experience felt that these streams were cool enough to sustain the fish, but the jury was still out on the experiment at the time the agency prepared its report.

Although the commission had established a game farm near Rogers and said the propagation work there was going "most satisfactorily," a shift in thinking was taking place nationwide within the conservation community. Arkansas's commission showed it was up-to-date, stating in its sixth biennial report that it hoped the 1927 General Assembly "will provide a proper system looking to the creation of state game refuges."

Just as the commission began to feel it was making progress—more with fish than game—it and the rest of Arkansas were devastated by three disasters in four years. The first was the Great Flood of 1927. The second was the stock market crash of October 1929 that marked the beginning of the Great Depression. This was followed by the drought of 1930, one of the three most severe droughts recorded in the state in the twentieth century.

As to the effects of the flood and drought, the commission described them vividly in its biennial reports. The seventh biennial report noted that when waters from the spring and summer floods of 1927 receded,

> millions of game and commercial fish were left stranded in the lowlands, and in the late summer and fall the U.S. Bureau of Fisheries joined this department in a cooperative plan for rescuing and reclaiming such of these fish as could be successfully handled.

In its ninth biennial report, the commission said,

> The historic flood of 1927, aided by increased artificial propagation and better protection, had built up a rather abundant supply of fish in many sections of

A few individual animals like this deer were rescued by well-meaning people, but the 1927 flood decimated wildlife throughout huge areas of Arkansas. Recovery was impeded by the Great Depression in 1929, then further crippled by the massive drought of 1930.

Arkansas Game and Fish Commission (below) *and Arkansas History Commission* (left).

Left, **Arkansas City, May 1, 1927**

Below, **Lassoing deer from boat**

the state. Game, particularly the larger types, was rapidly becoming rehabilitated in sections where the flood took its greatest toll and was steadily increasing in other territory.

The spring of 1930, however, to a great extent, destroyed what had been accomplished in fish propagation during the few years immediately preceding. Our rivers, as well as small streams, did not experience their usual spring flood stages. Lakes of the state did not overflow, and water levels, generally, were lower than for 50 years, if not in the history of the state. Spawning grounds for fish were reduced to a minimum, and the fish were forced into open waters, where they became easy prey for natural enemies as well as over-enthusiastic fishermen.

Literally millions of potential fish were destroyed. Great catches of fish, heavily laden with spawn, were made by thoughtless fishermen. The drought has continued throughout the spring and summer and, as a natural consequence, numerous bodies of water are drying up, while streams are reduced to occasional shallow pools. An additional heavy loss of fish is anticipated.

The drouth, with accompanying forest fires, reduced covers, and shortage of food, is working a great injury on our game resources, especially smaller upland types . . . bobwhite quail have doubtless suffered to a greater extent than any other species and it now looks that an alarming shortage is to be expected.

Budgetary shortfalls forced the closure of the game farm at Rogers as soon as the birds on hand could be disposed of and plans for taking care of the farm and equipment could be made. Construction work on Lonoke Hatchery was suspended.

Predatory animal control was reduced more than 50 percent, despite complaints from farmers, stockmen, and "those interested in game preservation." On September 1, 1931, the salaries of all employees drawing more than one hundred dollars a month were reduced 10 percent, after which "further adjustments and reductions" became necessary.

The commission confessed in its ninth biennial report that during these troubled times enforcement of regulations had been "extremely difficult." Some people "who were actually in need have taken both game and fish contrary to regulations in order to feed their families, while many unsportsmanlike persons have taken advantage of the economic situation and imposed upon the wild resources . . . both game and fish have suffered severely."

In 1932–33, the commission's income plummeted to $50,852.86 from $157,852.86 the year before. "If it hadn't been for a sizable cash balance from the previous year, it would have been necessary to eliminate practically all operations," Nell Taylor recalled in her 1957 report. The operating expenses in 1931–32 had been $153,692.47.

After this crunch, there was a gradual recovery until 1937 when the legislature, through the efforts of Chairman G. B. Oliver Jr., enacted a law increasing hunting and fishing license fees from $1.10 to $1.50 each. With this, the commission experienced a yearly income increase of about $100,000 until July 1, 1945, when the old commission was superseded by a new system put in place by voters through constitutional Amendment 35.

Though by no means well off, Arkansas had begun to edge out of the Great Depression in 1937, the same year the General Assembly passed Act 8, which abolished the old Game and Fish Commission and created a new one. It was in the next session that the General Assembly returned to its habit of enacting different wildlife laws for specific counties or groups of counties. The habit worsened in 1941 and 1943. There was a new twist to it, though. The Arkansas Supreme Court had ruled it unconstitutional for the General Assembly to enact "local legislation," that is, bills applying to only one county. To get around this ruling, the legislators began drawing bills with provisions that applied to counties within a certain population range, according to the last federal census. The population parameters usually were set so the law would apply to only one county. The effect, of course, was the same as having local legislation. The effect on the commission's ability to protect the state's game and fish was also the same.

Throughout their existence, the commission and the department had been vassals of the patronage system. This was true of every other state agency as well until the federal government demanded that a merit system be put into effect in Arkansas's new Welfare Department.

Before Orval Faubus in the 1950s and 1960s, it was a tradition (broken previously only by Jeff Davis) for Arkansas governors to serve two two-year terms. This total corresponded to the four-year terms accorded by law to Game and Fish commission members. Thus it was expected and tolerated that there was a complete change of personnel on the commission and in the department with every change of the state's administration. There was no stability or continuity in the management of the state's game and fish.

By the late 1930s, the state's sportsmen, led by the Arkansas Wildlife Federation, had had enough. The federation drafted a constitutional amendment designed to remove the commission from politics and obtained enough signatures to get the amendment on the 1940 general election ballot. The organization abandoned the amendment before the vote, however, because it was flawed: it failed to provide that the General Assembly would have to appropriate funds before the commission could spend them.

By October 1943, the Pulaski County chapter of the federation had drafted another proposed amendment and presented it to the commission. In May 1944, a federation committee asked the commission to give educational director Tom Mull a leave of absence so he could promote passage of the amendment. The commission refused, but this apparently did not mean the agency was opposed to the amendment. In fact, in September, the commission unanimously agreed to endorse the amendment but "expressed a desire that its support not be made a matter of general publication as this might lead to some opposition from the general public." In the general election of November 7, 1944, Amendment 35 was approved—115,214 in favor and 72,797 against—making the commission an "independent" state agency.

One of the ironies of the period surrounding this change occurred a year after the new commission had been installed. A member pointed out one new commissioner would be appointed each year. He feared that this might jeopardize the jobs of some good employees. The commission agreed unanimously that some internal provisions should be made so "worthwhile and capable employees should have some security and protection in their jobs." It adopted a resolution saying there were to

Amendment 35

Arkansas Game and Fish Commission.

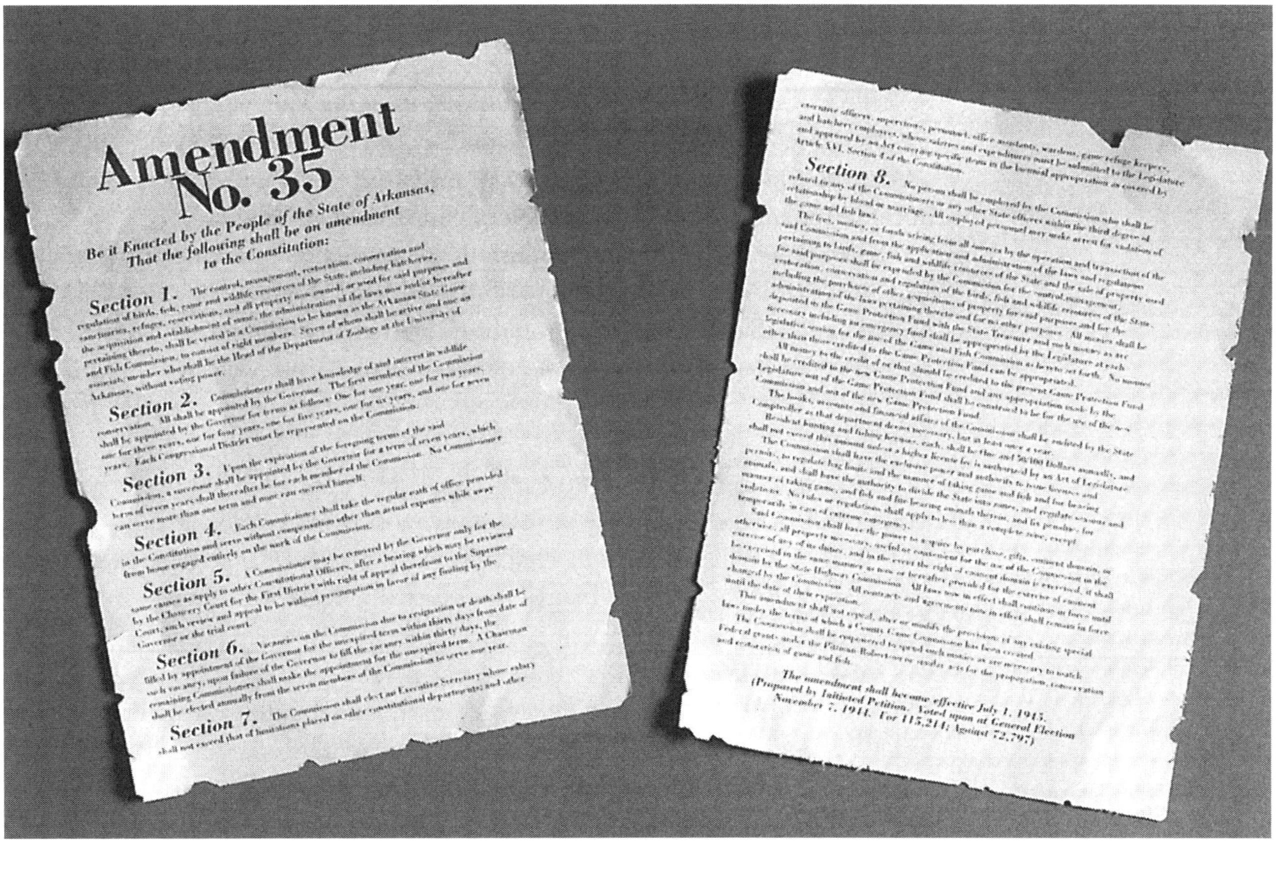

be no discharges "without just cause" and a majority of the commission, not just one member, would determine what was just cause. Further, vacancies were to be filled when possible from within the agency.

It was a new era, but the fact the commission would feel compelled to take these steps is an indication of the chaos prevailing until this point in Arkansas's conservation history.

NATIONAL DEVELOPMENTS

While states were struggling to pass game and fish protection legislation and establish administrative agencies, considerable activity was taking place at the federal level.

Rep. George Shiras III of Pennsylvania, a renowned wildlife photographer, was the first to promote the notion that migratory birds could be protected only by the federal government and not by the states to which the game "belonged" (Graham 1971). It occurred to him, while working on a bill designed to protect the public against polluted water served on railroad trains, that if the government could eradicate germs in interstate traffic, it could protect game birds that moved from state to state.

In 1906, Shiras introduced a bill in the House providing "that all wild geese, wild swans, brant, wild ducks, snipe, plover, woodcock, rail, wild pigeons, and all other migratory game birds which do not remain permanently within the borders of any State or Territory shall hereafter be deemed to be within the custody and protection of the Government of the United States."

Shiras's chief purpose was to have the Biological Survey (later the Bureau of Sport Fisheries and Wildlife) take over regulating the hunting of these birds. Sportsmen were beginning to realize spring shooting had decimated the flocks of wildfowl that once had seemed inexhaustible. Shiras was reflecting the sentiment of sportsmen who wanted hunting seasons restricted to fall and winter when young born during the summer had swollen the number of wildfowl.

States were not ready to relinquish their jurisdiction over migratory wildfowl, and Shiras's bill got nowhere. In 1912 after he left Congress, the great flocks continued to dwindle, and his legislation reappeared in much of its original form in bills sponsored by Rep. John W. Weeks of Massachusetts and Sen. George P. McLean of Connecticut. The bills' constitutionality was still questionable, and the passage of either appeared unlikely.

According to Frank Graham Jr., on September 18, 1912, a group of conservationists met in New York City to determine what could be done to salvage the legislation. By the time the meeting ended, McLean's bill had been broadened to cover all species of insectivorous and song birds that crossed national and state borders during migration. Including these species proved to be a shrewd political move, because the Biological Survey had been conducting a strong publicity campaign for years about the value of birds to agriculture. The rewritten bill reaped the benefit of that publicity, and farmers' representatives in Congress who had been apathetic about saving game birds were not going to get caught voting against "the farmer's feathered friends" and "destroyers of weeds and insects."

Both the Weeks and McLean bills slipped through committee hearings amidst much praise. The Senate passed the McLean bill without a dissenting vote, and it was added as a rider to the Agricultural Appropriations Bill. This bill reached the desk of President William Howard Taft on March 4, 1913 (Trefethen 1961). Taft signed what was known as the Weeks-McLean Migratory Bird Act without reading it because he was in a rush to wind up his duties before turning over the presidency to Woodrow Wilson. When a friend in New Haven later praised Taft for giving the country a federal law to protect migratory birds, the former president responded that he would have vetoed the entire appropriations bill had he realized that it contained what he considered an unconstitutional rider. The Weeks-McLean Migratory Bird Act had far-reaching effects. It dealt a blow to market hunting and made spring shooting illegal. It conferred on the secretary of agriculture the power to establish seasons on migratory birds. A few diehards refused to accept federal regulation. Rep. Frank W. Mondell of Wyoming, an ardent foe of federal regulation, introduced a bill to repeal the Weeks-McLean Act almost as soon as it passed. A test case was begun in the courts, and conservationists feared they were headed for defeat. However, Sen. Elihu Root, a former secretary of state, quietly had introduced a resolution authorizing the president to enter into international conventions to protect migratory birds even before the Weeks-McLean Act had passed. He saw the Weeks-McLean Act as unconstitutional and so tried an entirely new approach. His resolution, however, was introduced too close to the end of the session and saw no action.

On April 17, 1913, Senator McLean, with the Migratory Bird Act under heavy attack, introduced a new version of Root's resolution. It passed on July 7. Conservationists awaited the test court case outcome while negotiations for a Migratory Bird Treaty, complicated by the outbreak of World War I, were carried on between the United States and Canada. The document finally was signed in 1916

Woodcock with hatchlings

Migratory game birds such as the woodcock were hunted year-round prior to passage of the Weeks-McLean Migratory Bird Act in 1913. The act effectively ended market hunting of migratory birds and made spring shooting illegal.

Arkansas Game and Fish Commission.

and ratified by Congress in 1918. The Supreme Court then dismissed all challenges to the federal government's jurisdiction over migratory birds because the provisions of a treaty signed with another nation take precedence over the rights of individual states (Graham 1971).

From that time, Arkansans had to abide with U.S. Department of Agriculture regulations governing migratory birds, but it was not until 1931 that the state legislature formally relented and acknowledged federal preeminence by adopting Act 86. This law stated that when federal regulations governing the hunting and taking of migratory birds were changed, the Arkansas Game and Fish Commission could make state regulations conform to the federal rules. Any regulation passed "shall be the law of this State," the General Assembly declared, and violation was a misdemeanor punishable on conviction with a fine of not less than twenty-five dollars. This was by no means the extent of federal activity during the early 1900s, but it was the one area that established precedent.

9 The Conservation Debate

Even after market hunting was under control and an agency was established to administer the state's game and fish laws, most Arkansas wildlife populations continued declining dramatically. Conservationists searched diligently for causes and came up with many. Some focused their anger on a few causes and a single solution. Witness W. S. Allen in the *Arkansas Deer* (1924):

> Now, when one goes into the forest, a stillness reigns that makes men of those days feel a lonesomeness and sadness that is hard to express. It is though nature had deserted its accustomed haunts . . . They [the fowl] are now all practically gone and those that remain seem to have a constant fear that they too are soon to go the way of their forbears[*sic*]. It is all due to modern firearms, modern dogs (I never heard of a bird dog until I was long past man's estate), settling up the forests, the fur hunter and a reckless disregard to our wild life. The only remedy, . . . to pass a law forbidding all hunting for a period of at least five years.

In blaming hunting and focusing attention primarily on weapons, the writer was following the national lead set by New York Zoological Park director Dr. William R. Hornaday. One of his favorite targets was the autoloading shotgun. The autoloader, improperly used, was a deadly weapon, but no more deadly in the hands of a skilled shooter than the pump gun (Trefethen 1961).

Others saw the problem as extending far beyond the hunter to the civilization that Americans had chosen to build for themselves.

In the *Arkansas Conservationist* in 1926, Dr. J. C. Futrall, former president of the University of Arkansas, spoke to the changes in the way people were living and what it was doing to his favorite game, the quail:

> Fifteen years ago there seemed little danger that quail would at any time soon become even scarce in the state . . . In many parts of the state the clearing away of patches of timber and brush on the farms has destroyed much of the cover that formerly protected the birds. Accompanied by a friend, and with two excellent dogs, I spent several hours on a good day hunting over excellent territory,

Cleared timber

Major clearing of Arkansas's forests severely degraded wildlife habitat. During a flourishing period at the height of World War I, the logging industry consumed much of the state's forests.

Arkansas Game and Fish Commission.

> and found only six birds. My friend told me that three years ago he had hunted exactly the same territory and had found 14 coveys in one day.

Dr. Futrall's answer was to shorten the season and lower the bag limit. "If something of this kind is not done, and done quickly," he warned, "Arkansas will soon be a state where it will hardly be worthwhile for one to go into the field with a gun."

The Arkansas Game and Fish Commission, citing support from the *Arkansas Democrat,* was promoting more hunter restrictions. "In the opinion of the *Democrat* an absolutely closed season should be put into effect to protect the dwindling number of deer and wild turkey," the commission's magazine reported. "If this is not feasible, then at least a law should be enacted prohibiting the use of hounds in the hunting of deer."

Why the commission saw hunter restrictions as the first line of defense was explained this way in the *Arkansas Conservationist* (1, no. 4, 1924):

> A few years ago but few hunters obtained the daily bag limit on more than one day during the season. It was impossible for the hunters to get into the sections where game abounded, except when the most favorable weather conditions prevailed. The building of good roads into the hunting grounds, however, has made it possible for hunters to go into the fields almost daily.

Trying to achieve control through higher license fees was unacceptable to the commission because this unfairly affected lower-income hunters. That, the agency concluded, means the "only solution is to radically reduce the bag limits, shorten the hunting seasons, and strictly enforce the game protection laws."

The commission saw restoration as another of its primary duties. "It is difficult to conserve or preserve game where the supply is almost entirely exhausted. On the other hand, judicious propagation and distribution will soon increase the stock and replenish the depleted coverts." As indicated, the commission turned first to the idea of game farms and fish hatcheries to carry out its restoration plan.

The agency was coming under heavy attack for the continuing disappearance of Arkansas's wildlife, as evidenced by this defensive editorial statement in the *Arkansas Conservationist* (1, no. 4, 1924):

> By all that is fair and just, the Commission cannot be rightfully censured for not getting desired results until the legislature has created laws out of its recommendations and a thorough trial has been given the proposed system. If we are going to have a Commission, its proposals should be given consideration because the people expect appreciable results; [if not], why not do away with the department and get started on the right road?

Observers with a broader perspective of the overall problem offered more varied solutions. Lance Hemingway, speaking to the game situation in the *Arkansas Conservationist* (1, no. 1, 1924), wrote that the disappearance of Arkansas's wildlife

> kept pace with advancing civilization for years, and latterly outstripped it altogether. With the coming of modern firearms—all of them improvements over the pioneer's muzzle-loading rifle—it became a mere matter of getting to the game country in order to get the game. An ever widening agriculture drove the game from the more open lands to the mountains and swamps. Constant hunting with dogs thinned the deer population and drove the pitiful remnant to remote and inaccessible places in swamp or mountain fastness.
>
> Railroads running in every direction offered quick and easy access to the remoter sections of the State, and parties of "camp hunters" helped to decimate the already pitiful numbers of deer and bear which sought refuge in swamp and cane brake. Armed with repeating rifles and double-barrel breech loading shotguns, these hunters who used the railroad to reach their quarry, made distressing inroads in the game supply, but the equipment of the latter part of the 19th Century was insignificant compared to the equipment of today. Armed with high-power repeating rifles and automatic shotguns, hunting parties now speed toward distant hunting grounds in automobiles, which are capable of penetrating wherever there is even the semblance of a road, except in extremely bad weather. The toll . . . destroyed the deer in county after county and almost destroyed the bear from one end of the State to the other.

Hemingway urged Arkansans to look beyond their state's borders for solutions:

> With the examples of New York, Pennsylvania and other of the older states before us, we need not fear the results if we protect our wild life as it should be protected. By creating closed seasons, fixing bag limits, forbidding the killing of female deer and wild turkey hens, and by other means, these older states have succeeded in augmenting their game supplies and wild life is now plentiful where it had been almost completely destroyed by constant and indiscriminate hunting. Arkansas now forbids the killing of female deer and turkey hens, and some other needed precautions have been taken, but [lack of other adequate legislation] is still apparent, and these assets are gradually disappearing from whole counties where they were once abundant.

"Carelessness of consequences," Hemingway continued, "has almost exterminated the prairie chicken and wild turkeys, and only our oldest citizens remember the Carolina parakeet, which as late as 40 years ago had not wholly disappeared from the Mississippi Valley, and which, prior to that time, was plentiful as far north as St. Louis."

Camp Lyronza, November 24, 1904

A typical turn-of-the-century Arkansas hunting camp.

Arkansas History Commission.

He lamented the loss of another creature, the sandhill crane, which he said was "fairly common in Arkansas" until it fell victim to man's lust for slaughter. Hemingway asked, "Why was he hunted and slain by Tom, Dick and Harry? Not for food because there were other birds whose flesh was more palatable. Not for sport because there is no sport in shooting a defenseless bird or animal which has no value as food or a furbearer. Not for plumes because these were not of great value and the market was uncertain. The impelling motive, then, must have been a mere lust for slaughter."

In the *Arkansas Deer* (1924), Hemingway warned the fate of extinction was awaiting the bobwhite quail, but he did not blame its suffering on man's lust for slaughter. His culprit was the farmer:

> Farmers, in clearing their lands and putting their farms in condition, cleaned the fence corners and cut away the briers and shrubbery from beside creeks and ditches. Thus cover needed by the quail for protection against natural enemies and cold weather, and for breeding places in summer, w[as] destroyed.
>
> No farmer actually needs the ground embraced within the corners of his worn fence, or the few feet of space alongside the modern wire fence; nor does he plow right up to the edge of creek or ditch, notwithstanding the space has been cleared of underbrush and weeds. How much better it would be if he would leave cover for quail and other useful birds along his creeks and ditches . . .

Not every species was gone, and not every location in Arkansas was devoid of sport as a result, based on evidence from Jo Ann Browning, author of a history of St. Charles, a small city on the lower White River. "St. Charles," Browning said, "was first recognized as a hunter's paradise about 1930. Duck and deer hunters began to pour in and they have not stopped yet [1953]. There is quite a bit of excitement when the deer hunters go down the river to their camps and quite a bit of sleep lost as they leave at the crack of dawn to go duck hunting."

Just as many causes were being advanced in the 1920s for the shortage of fish in Arkansas's waterways. But the fish problem was even more complex than the game problem because conditions varied considerably from stream to stream.

H. S. Davis, a pathologist with the U.S. Bureau of Fisheries, told in 1924 of visiting the Arkadelphia area "to investigate the cause of the notable scarcity of game fishes in the Ouachita River and its tributaries." The bureau had received reports that pollution was the cause, but Davis said it did not take more than a cursory exam on his part to determine that pollution was not the culprit. It was true, however, that only a few years earlier, both the Ouachita River and its tributary, the Caddo River, were famous for their bass. "It was a poor fisherman indeed who could not show a good string after a few hours' sport along the Ouachita or the Caddo. The former furnished ideal conditions for the large-mouth bass, while the swift waters of the Caddo were equally well adapted to the small-mouth. But times have changed, and it now requires an exceptionally lucky or skillful fisherman to catch two or three good bass . . . [in] a day's fishing."

Davis said two major factors caused the "wholesale depletion" of game fish in the United States. They were "the changed environmental conditions . . . and the enormous increase in the number of anglers." The Ouachita and the Caddo were still "ideal bass streams in spite of the fact that the fish are not there." So the cause had to rest with the second fundamental change—overfishing. An angered Davis went on to ask, "How can the people of Arkansas expect game fish to flourish in their streams when there are practically no restrictions on fishing and any man can catch as many fish as he is able, at any time of the year, and in any manner he may choose to employ?"

D. G. Beauchamp, writing in a 1926 issue of the commission's magazine, had no trouble discerning what caused the demise of fishing in the St. Francis River, which was once proclaimed as "the best place in America to fish for black bass": "Many lay our loss to the ditching, while others blame the netting, trapping and seining. While these have been efficient causes to bring about a deficiency of the finny tribe, they are not the principal cause as to the St. Francis River."

Bass anglers, Black River

Bass anglers with a catch near Corning on August 7, 1918. Several Arkansas rivers—the Black, Ouachita, Caddo, and St. Francis, to name a few—were considered among the country's best black bass waters in the early 1900s. Catch rates declined by the 1920s, however, in part because of unrestricted fishing.

Arkansas Game and Fish Commission.

In the *Arkansas Conservationist* (3, no. 3, 1926), he asserted that the major cause affecting the St. Francis River lay in the fact that

> certain ditches have been dug for the purpose of building levees between which it was hoped to hold the waters when in flood stage. Between these levees is the river, which divided into lakes, ponds and runs between which were many acres of smartweed beds, flags and willows.
>
> This space between the levees is waste and is never intended to be cultivated . . . Prior to the levee ditches this space was covered with water. The water from the river has left its former places, going into the ditches, leaving the space between the levees to dry up and destroy the breeding and raising place for the fish . . . The smartweed beds are what made the St. Francis the best fishing in the world, at one time. Now, if we will restore the water to the old river runs it will not be long until we will have lots of good fishing for all.

Guy Amsler, editor of the commission's magazine, acknowledged that causes varied from stream to stream. However, if the issue were limited to black bass alone, Amsler said he would have to point an accusing finger at the artificial baiter. In the *Arkansas Conservationist* (3, no. 1, 1926), he cited an unidentified lake in southeast Arkansas, some seven miles long, that he said had no rough fish or gar and had never been seined or netted. "It does not overflow and has never been drained. Dynamiting, shooting and gigging are unknown. In fact, few, if any, fish have been caught from the lake except with hook and line," he asserted:

> Many is the day that an individual bait fisherman has taken well nigh 100 fine bass from this lake and thought nothing of it . . . Female bass filled to the overflowing with eggs have been taken by the countless thousands. The worm and live minnow have been replaced by the "wooden monstrosity" with 15 and 18 hooks while bait and fly fishermen have increased a hundred and more times over.

Other streams and lakes were depleted for other reasons, Amsler acknowledged. He told readers that all reasons "have played a more or less conspicuous part in the extermination," and that various prejudices should be put aside when considering regulations and "each practice" should be examined for "its reasonable share of responsibility . . . Our fish have gone, no one questions this, and the big problem now is to bring them back."

As with game, Lance Hemingway (*Arkansas Deer,* 1924) saw what was happening to Arkansas's fish earlier than most:

> Forests were cleared for cultivation, and water courses which had been running streams all the year round felt the effect of the denuded lands drained by them. Waters formerly held in check by vegetation rushed down the slopes, filled the streams to overflowing, and hastened to the sea. Later the streams dwindled to stagnant pools, and the fishes in them suffered accordingly. Creeks, and even the "branches," which were formerly full of fish in spring and summer, were no longer productive, and the rivers and lakes became the resorts of fishermen. Soon these felt the effect of indiscriminate seining and netting, and later the dynamiter, the "lime jugger," and the use of India berries completed the destruction.

Harry Walsh, a Canadian waterfowl biologist, summed up the situation in an article reprinted in *Arkansas Game and Fish* (fall/winter 1975):

> We share a common environment with other members of the animal kingdom. If we destroy their habitat, we also destroy our own. Earth, our home, sits like a jewel in the center of infinity. Ours is the only planet capable of sustaining life as we know it; yet man is unable to get along with himself and members of his own race or to share with others of the animal community. You cannot help but wonder what chance the poor old duck has.

What chance indeed? It also might be asked what chance the Arkansas Game and Fish Commission had in light of what was occurring in the state during the 1920s and 1930s.

As previously noted, lumber production in the South peaked in 1909. During World War I, the expanded need for lumber increased logging of southern pine forests. This was a flourishing period of southern lumbering, and it brought unaccustomed prosperity to established towns and new backwood communities. A broader tax base brought new roads, new schools, new courthouses, and other civic improvements. The hungry industry swallowed forests insatiably. The pineries, once thought to be inexhaustible like prairie chickens and fish, began to vanish. By the late 1920s, lumber production declined. Across the South, officials watched the clearing of timber and the closing of mills. Many operators with timber mortgages and other debts to pay simply ran their mills day and night, some without profit, in order to liquidate their forest holdings quickly. During the 1930s, many companies abandoned their harvested woodlands and let the counties take them for nonpayment of taxes (Clepper 1971).

Remaining stands of timber and cut-over land were burned. As soon as the areas had been set aside by President Theodore Roosevelt, rangers in the Ouachita and Ozark National Forests began arduous efforts to persuade Arkansans not to burn the woods. But it was not until 1927 that the American Forestry Association began an ambitious three-year project designed to change the tendency of rural people to burn the woods every year (Clepper 1971).

The Great Flood of 1927 wreaked economic havoc in the Delta. Levees built previously were breached; some were just swept away by the force of rolling walls of water.

According to *A Survey of Arkansas Game,* written for the commission in 1951, the St. Francis River area had been converted from a malarial swamp into a rich farming area. Construction of drainage projects started in the lower St. Francis River basin about 1919. By 1930, 59 percent (470,000 acres) of this area had been improved through drainage. Between 1930 and 1940, reclaimed lands increased at the average annual rate of about 13,550 acres, much of it through work of the U.S. Army Corps of Engineers. The rate rose to 16,000 acres a year between 1940 and 1945, and by 1947, about 89 percent of the area—685,000 acres—had been drained. Except in extraordinary years, the threat of another great flood was eliminated.

The Arkansas Game and Fish Commission's recognition of the importance of habitat had grown by 1927. The agency said there was no greater winter population of mallard ducks on the American continent than in eastern Arkansas. In the state's rice belt, ducks found unlimited food in fields and overflow areas, while countless thousands of wild, watered acres provided resting places. However, the commission said in its seventh biennial report,

Excavation of drainage ditch on St. Francis River

The St. Francis River became a classic example of man's attempts to control nature. Drainage projects in the lower basin began in 1919. By 1930, 59 percent or 470,000 acres were cleared and drained. By 1947, the figure rose to 89 percent or 685,000 acres drained. Similar projects throughout the Delta destroyed valuable wetlands habitat and its wildlife in order to create agricultural land. Construction of drainage ditches often resulted in rich bottomland hardwood habitat being cleared of trees and planted to cotton.

Arkansas Game and Fish Commission.

> Hunters in annually increasing numbers, good roads, and motor boats at greatly reduced prices, are gradually bringing heretofore inaccessible shooting areas within easy reach of the gunner. Drainage has also destroyed a large number of excellent natural preserves. If, therefore, present plans for a nationally administered flood control project should materialize and further reduce the state's overflow district, we would immediately be faced with a waterfowl food and resting grounds problem. Just how this problem can best be solved is subject to divergent views. It is exceedingly important, however, that the state give some thought to this situation and at the proper time act in conjunction with the Federal Government, or independently, if necessary.

International and national projects were driving home the importance of habitat to society's thinkers and doers. It was rapidly becoming apparent that it did little good to stop spring killing of migratory fowl under the Weeks-McLean Law if the birds were being deprived of places to rear their young after arriving in the north.

Planting cotton with mules in Poinsett County, May 1943

The commission noted in its 1923 report that "the flight of ducks and geese today [over Arkansas] is much greater than it was eight years ago." It went on to say the increase was "due to the untiring efforts of a coterie of sportsmen scattered throughout the United States, with little or no assistance from the shooters at large."

The International Association of Game and Fish Commissioners, consisting of American and Canadian wildlife managers, was making arrangements to protect sufficient breeding stock at the end of each year's shooting season and had persuaded the Canadian government to buy "large breeding grounds" that were being protected. Before the land was purchased, the commissioners explained in the fourth biennial report, "pot hunters and game hogs would go into this territory, rob nests and slaughter the young, so that it was impossible for these birds to continue to produce the enormous flights seen by the Southern shooters." Further, a fund was being collected to purchase a winter home for the fowl on the coast of Louisiana: "The United States can boast of this winter home, extending over a frontage of more than 70 miles of the Gulf, ranging back about 20 miles in the interior—ducks and geese so wild that one cannot approach within gun-shot outside of this reservation, will feed from grain that you may pitch to them on the reservation, showing how soon the bird learns where he is protected and where he is not protected."

The association also pointed out Congress was then considering what is known as the "Public Shooting Ground Game Refuge Bill" providing for a general federal hunting license to shoot migratory birds with the fees collected to go into a fund to buy public hunting grounds and refuges for migratory birds and other game.

Arkansans needed to support this bill, the Game and Fish Commission warned because "all of the best shooting territory [in Arkansas] is taken up by private clubs. If this continues, it will be a matter of a short time that nowhere in Arkansas will there be any good shooting territory left open to the farmer boy without means."

The establishment of Big Lake National Wildlife Refuge in Mississippi County in 1915 as the first national wildlife refuge in Arkansas—and one of the earliest in the country—was noted earlier. Some ten years later, efforts in Arkansas's national forests began to shift into resource development and improvement. In the Ozarks, this began with establishment of four federal game refuges by proclamation of President Calvin Coolidge. Unlike other government pronouncements, this action had great popular appeal, especially to local sportsmen. The Game and Fish Commission stated in its sixth biennial report that it was "co-operating with the Forest Service to the fullest extent in restocking these areas." The refuges succeeded and led to the eventual establishment of state and federal wildlife management areas within both national forests. By the fall season of 1940, the Sylamore Ranger District alone counted three thousand hunters. The deer herd in Stone and Baxter Counties

Erecting a game refuge sign

In 1929, the Game and Fish Commission began setting aside preserves meant to supply surrounding lands with game. By 1960, the commission was operating twenty-two game refuges covering 725,520 acres. In later years, biologists realized that areas closed to hunting produced no more game over the long term than other areas where regulated hunting was allowed. Habitat was the real key to preservation.

Arkansas Game and Fish Commission.

Wild turkeys on Howard County Game Refuge, winter 1941–42

This thirty-three-thousand-acre refuge, established in August 1930, provided many turkeys, deer, and beavers for stocking in parts of Arkansas where game stocks had been depleted.

Arkansas Game and Fish Commission.

increased under intensive wildlife management from twenty-four deer in 1924 to more than three thousand deer in 1942.

Meanwhile, more and more was learned and published about game refuges. For example, *American Field* magazine warned its readers in 1926 that it was not enough to protect game placed on refuges. Animals had to be managed so they would be protected from each other. The writer explained that "three classes or groups of animal life" existed: the first typified by the rabbit and game birds; the second or "middle class" exemplified by the fox, raccoon, skunk, coyote, opossum, and others; and the third class of larger carnivores such as the cougar, lynx, and wolves. It was essential to study the relationship between these three classes and understand how they act as checks on one other if a refuge was to succeed. The article then described that relationship.

Encouraged by preliminary results coming from forest game refuges, the Arkansas Game and Fish Commission began a program of its own with the help of local sportsmen, one example of which was described in "Howard County Game Refuge and Deer Farm" by Stella Wynn: In 1929, for example, John Kirby, A. L. Keith, and Lee Ward wanted to do something about rapidly diminishing game in northern Howard County. They contacted the commission in early 1930, and W. N. Deaton, supervisor of state game refuges, was directed to survey the area. He went to Howard County to investigate, checking food conditions, cover, the water supply, and the amount of game remaining. Deaton told the commission the natural set-up looked favorable, except little wildlife remained. There were, perhaps, twelve or fifteen deer and twenty or thirty wild turkeys. It was a well-wooded area with bountiful crops of wildlife food, was bordered on the west by the Cossatot River, and had Camp Creek, Snake Creek, Harris Creek, and other streams tumbling through it, along with several springs.

Refuge work was just beginning in Arkansas. Three refuges had been established, and the commission wanted to be reasonably sure this venture would succeed. It finally was decided to establish the refuge, and the matter was taken to area residents. Some viewed the idea with skepticism, but the Dierks Lumber and Coal Company (now Weyerhaeuser Company) favored the idea and helped substantially.

An ownership map was prepared covering thirty-two thousand acres in the center of the best wildlife area, the region where the remaining game animals were making their last stand. Petitions were drawn up, signed by landowners, and approved by the commission. The owners leased the property to the commission for ten years; the leases later were renewed for fifteen years and embraced thirty-three thousand acres. The original refuge was dedicated on August 14, 1930.

The refuge soon claimed the distinction of having the only thriving beaver colonies in Arkansas. Ten beavers were stocked on Snake Creek about 1932. They multiplied rapidly and dispersed over the refuge and outside it. On February 11, 1932, a tree felled by beavers in the area was presented by Deaton to the University of Arkansas museum. Also, deer in the area "increased phenomenally." The refuge contained the only "deer farm" in the state, in part because Dierks Lumber and Coal Company gave the commission an easement on 720 acres without charge, and because the U.S. Fish and Wildlife Service approved the plan, making it eligible to receive 75 percent federal funding under the Pittman-Robertson Act. Ten to twelve deer were placed within a new eight-foot fence made of specially woven wire. In addition, the Harvey Couch deer herd near Hot Springs and other deer were purchased until the breeding herd totaled fifty-six. By the late 1940s and early 1950s, deer were so plentiful they were being trapped and transported to restock other areas.

There were twelve state game refuges at the end of 1932. Four refuges had been returned to unprotected status after five years. "Public sentiment in the state favorable to the establishment and maintenance of refuges and upland game has increased by leaps and bounds during recent years," the commission told the legislature in 1934. "The few refuges in existence have demonstrated beyond question the importance of a state-owned and administered system of game refuges in a well-rounded game management program."

The Pittman-Robertson Act was one of two legislative conservation milestones enacted by Congress during the Great Depression. The first was the Duck Stamp Act, a direct result of the dust clouds boiling across the drought-stricken Midwest in 1929. The drought brought a new, immediate crisis to waterfowl (Trefethen 1961). Ducks returning to breeding grounds in the northern plains states and prairie provinces of Canada found ponds, sloughs, and potholes empty. A Bureau of Biological Survey study revealed prompt action was needed if waterfowl were to be preserved for future generations. The bag limit, as of December 31, 1929, was cut from twenty-five to fifteen ducks a day to compensate for damaged habitat and the resulting loss of young birds. Waterfowl populations continued declining.

Two half-forgotten, old proposals were resurrected before the Senate Special Committee on the Conservation of Wildlife Resources: one advocated a federal hunting stamp to finance waterfowl programs; the other advanced a one-cent tax on each shotshell. On April 4, 1932, the special committee heard more than one hundred witnesses, a majority of whom favored the stamp proposal, feeling upland game hunters should not be taxed to finance waterfowl management programs. With the air cleared, progress toward enactment was rapid, and the Migratory Bird Hunting Stamp Act became law on March 16, 1934 (Trefethen 1961).

First federal waterfowl stamp, by J. N. "Ding" Darling

Darling's mallard scene was the first of the annual series of migratory waterfowl hunting stamps issued by the federal government. It was released in 1934. Proceeds from sales of the stamps are used to buy and maintain national wildlife refuges administered by the U.S. Fish and Wildlife Service.

U.S. Fish and Wildlife Service.

Under continuing pressure from hunters and conservationists, Congress reluctantly began to include wildlife welfare in its deliberations. It passed the Pittman Robertson Act in 1937, calling for a 10 percent excise tax on sporting arms and ammunition with the proceeds going to wildlife restoration projects nationwide. Congress was much slower to react to depletion of the best fishing waters in the country. The dam-building craze that hit Arkansas and the rest of the country often turned swift-running streams into chains of placid lakes. Industrial pollution, domestic waste, and pesticide runoff transformed once-pure fishing streams into sewers. But it was not until 1952 that Congress passed the Dingell-Johnson Act, a counterpart to the Pittman-Robertson Act, which put an excise tax on fishing tackle to raise funds for fisheries research and restoration (Graham 1971).

Between the time Congress passed what has become known simply as the Duck Stamp Law and the Pittman-Robertson Act, Arkansas gained its second and perhaps best known federal refuge, White River National Wildlife Refuge. Established in 1935, the refuge encompassed nearly 113,000 acres. Initial purchases were made primarily from large timber companies. Most of the area was logged, and few standing hardwood trees remained. Game populations were depleted, and some questioned the wisdom of buying a huge parcel of decimated land. But later, Civilian Conservation Corps personnel developed the area by planting trees and by constructing fences, roads, water-control structures, and administrative buildings. As time passed, the scars healed. The forests grew back, and game flourished once again. Waterfowl were the primary game animals fostered by the refuge; but deer, turkeys, and other game thrived as well (Kaffka 1971). In fact, deer and turkey populations grew to numbers second to none anywhere in the state, along with one of the finest populations of small game, particularly squirrels, rabbits, and raccoons. The refuge's large size made it capable of accommodating a huge number of hunters.

With established refuges, the wildlife management learning curve rose rapidly as

the 1930s ended. Some of what Harold Alexander, dean of the Arkansas conservation movement, wrote in a fall 1976 article for *Arkansas Game and Fish* would be just as applicable to the situation that existed in the late 1930s. Alexander said, in part,

> Much of our recently acquired knowledge has, in a sense, been used to limit or prevent unwise courses of action, which have changed with new circumstances. This is demonstrated by the buck law, once valid, but now a detriment to good practices . . . On the other hand, we have not even begun to solve some of the major problems confronting us, and in some directions we seem to be progressing backward. Each day we are confronted with the accelerated impact of increasing human use of resources and particularly land use, for practical profit motives, which serve single objectives or the interest of a few; or are dedicated to "multiple use" purposes, meaning their dedication to the interests of a particular group or interest . . .

White-tailed deer

This herd was photographed on the White River National Wildlife Refuge in 1993. Established in 1935, White River was the second national wildlife refuge in Arkansas. Though game populations were depleted and most of the area had been logged at the time the land was purchased, this huge tract of bottomland hardwood forest quickly healed and soon supported impressive numbers of deer, turkeys, black bears, waterfowl, and small game animals.

A. C. Haralson, Arkansas Department of Parks and Tourism.

10 Status of the Species, 1900–1945

At the end of its first year as an administrative department of state government, the commission summarized that, among the species of "greatest importance," Arkansas had "a few bear, fewer beavers, some otter, prairie hens, wild turkeys and white-tailed deer." There were enough of these species left, the agency added, that if properly protected, they could replenish the forests in "great numbers." At the time, however, "the annual slaughter is greatly in excess of the annual increase bred, and this can lead to but one result—annihilation."

The state's principal game birds in 1915 were turkeys, quail, and prairie chickens. The commission also made a valiant effort to stock the state with ring-necked pheasants. The prairie chicken had almost been exterminated and eventually was annihilated, partly because protection and adequate enforcement came too late, but mostly because of continued habitat destruction.

UPLAND GAME BIRDS

WILD TURKEY

Several factors affected turkey populations between 1900 and 1945. Arkansas's original turkey flocks suffered heavily from cutting of virgin timber, but in some areas, turkeys were gone before the trees were cut. The birds became concentrated as civilization spread, making them targets for illegal hunting and overhunting. In addition, turkeys were hunted for the market until 1915, even though a decrease in their population was seen as early as 1900.

The commission reported in its sixth biennial report that at the Rogers State Game Farm, a "very good bunch of wild turkeys" was being built, so "propagation in considerable numbers can begin immediately on the passage of necessary protective laws." Despite these efforts, *A Survey of Arkansas Game* (1951) reports the turkey population reached a low in 1930, a year of fierce drought. The spring 1971 issue *Arkansas Game and Fish* said the population was thought to "have reached its low in the 1920s and early 1930s when the flocks dwindled to insignificance in many parts of the state and that the only areas where they were relatively plentiful were in isolated places along the Mississippi and the lower White and Arkansas rivers."

In the mid-1930s, the commission said there were conflicting reports about the size of the state's turkey population. Some said the supply was decreasing gradually, the commission noted while others insist it was increasing. In some sections, at least, it said, "there is undoubtedly an improvement, and turkeys were killed during the biennium in some counties where there had been none for a good many years." Turkey hunting, it added, had improved "wonderfully" in the vicinity of game refuges. Whereas in 1915 there was a four-turkey bag limit, the commission said in the mid-1930s "any hunter should be satisfied with one gobbler per year, regardless of whether he is killed in the fall or in the spring."

A major restocking effort began in 1932, the same year the game farm's operation was discontinued. In 1972 Bill Gaines in *Arkansas Game and Fish* said the commission realized the best procedure in a restoration program was to transport native wild turkeys trapped in one location in the state to a new location. But with few employees, a small budget and the physical difficulty involved in trapping, this was easier said than done. As a result, thirty-one turkeys were acquired from North Dakota and Mississippi and released in refuges in Howard, Ouachita, Grant, and Monroe Counties (Holder 1951).

Five hybrid turkeys were released near Galena, Howard County, in 1934 and reportedly thrived. The first efforts to trap turkeys in Arkansas were made in 1937 when 11 were taken from the Howard County refuge and released in Monroe County. From 1937 to 1960, turkeys were trapped in limited numbers and stocked around the state. The demand for turkeys to stock was much greater than could be met (Gaines 1972). With federal funding, the commission obtained 210 pen-raised wild turkeys from a commercial game farm in Missouri in 1940 and 484 in 1941. Groups of 30 to 40 were released in White, Lafayette, Jefferson, Drew, Columbia, Union, Pike, Stone, Calhoun, Boone, Howard, Prairie, Lee, Grant, Clark, Montgomery, Franklin, and Pulaski Counties.

After being released, the turkeys were held in open-topped pens a few days to localize them and prevent their wandering or drifting. One source said after being released, the birds immediately drifted away, and some moved as far as fifteen miles (Holder 1951). Another reported, however, these game farm birds, while they originally came from a wild strain, were several generations removed from the wild and had lost their inherent wildness. They were so tame they tried to follow people (Gaines 1972). Some commission employees also saw a "very definite danger" the turkeys would introduce disease into the existing wild population. Although the writer of a 1972 article in the commission's magazine conceded, "The success of the

Left, **Turkey release**

Below, **Turkeys in zoo pen**

pen-raised turkeys is not known," it was assumed they "made delicious coyote and bobcat dinners. Nevertheless, the same error was repeated again in 1958 with the release of 200 more pen-raised birds." More than one thousand eggs laid by the confined birds in the first shipment were collected and sent to the Windy Willow Hatchery, where the poults were raised to the age of ten weeks and returned to the commission for restocking efforts.

Drew County had an estimated 261 turkeys when it was surveyed by the commission in 1944, and local residents believed half were offspring of birds released in 1941. Other south Arkansas counties reported some of their flocks were descendants of the 1941 restocking program, but the project as a whole was termed a failure because some stockings in Stone, Franklin, and White Counties resulted in the total loss of all the birds released. More success was achieved where at least a few wild turkeys existed before the release; the poorest results were in areas where native turkeys had been exterminated.

The total reported turkey kill in Arkansas was 202 in 1938 and 299 in 1945.

The Game and Fish Commission began trapping and relocating wild turkeys in 1932 to help reestablish flocks in low population areas. In September 1941, this flock was being held at the Little Rock Zoo until time for its release.

Arkansas Game and Fish Commission.

NORTHERN BOBWHITE

Quail were more abundant than turkeys in 1915, the commission reported, but there also was "greater temptation to shoot them." The agency concluded there were enough quail to be hunted freely, provided they were "taken care of during the closed season."

A few quail were raised at the Rogers State Game Farm. Some were stocked on refuges created by groups of farmers. Most quail from the game farm were exchanged with other states for species or eggs of species Arkansas wanted.

Right, **Quail demonstration area in Faulkner County**

Below, **Releasing quail**

An intensive quail management program began in Arkansas during the Great Depression and continued through the 1940s. Demonstration projects were set up in twenty-four counties, including one shown here in Faulkner County in 1943 where a mule-drawn harrow is being used to prepare a lespedeza border strip. Surplus bobwhites on some of these areas were trapped for restocking.

Arkansas Game and Fish Commission.

As tight as money was during the Depression, the commission still opted to start one new program, two three-year quail habitat demonstration projects to show how to provide proper cover and food. One project was on 960 acres near Roe (Monroe County); the other was on 1,305 acres near Monticello (Drew County). What the commission hoped to gain from the demonstration was information it could pass on to landowners about increasing the number of quail on their property and the cost. Meanwhile, the commission said the quail population in the previous two years had shown a "splendid increase."

Two years later, the commission, according to T. H. Holder in *A Survey of Arkansas Game,* declared the experiments successful. Information revealed that "quail respond more readily and rapidly to proven game management methods than any species in the state and increasing the supply as a rule involves less expenses and less time than is required with species that do not reproduce as rapidly."

A federally aided demonstration project patterned after the experiments began in 1940. It was modified to provide for trapping surplus birds for restocking. Demonstration projects were set up in twenty-four counties on areas covering two to three thousand acres and were closed to hunters for five years. Habitat improvement increased the number of quail.

The most intensive quail management in Arkansas was done on a private shooting preserve near Roe, bought for that purpose in 1939. The preserve contained about three thousand acres, mostly farmland broken by patches of hardwood forests, scattered thickets, and old home sites. Quail increased rapidly the first three years and peaked in the winter of 1941 when 250 coveys were reported. Although quail management practices continued, the population began declining, and the owner wrote in 1945 that he doubted if there were as many as 250 birds in the entire preserve. Statewide, however, quail populations appeared to be fairly stable because many hunters reported killing their limit on the last day of the sixty-one-day season in 1945–46.

Arkansas's quail refuge program won national attention. In one form or another, the Arkansas plan was put into practice in most states where quail are important. The plan called for three or more farmers with adjacent property to agree to cooperate for five years in providing favorable environment for the bobwhite and to prohibit hunting. "The result has been an overflow of quail into adjacent territory [and] has produced excellent hunting with a minimum of expense and effort" (Black 1978).

RING-NECKED PHEASANT

Ring-necked pheasants were the principal product of the Rogers State Game Farm. In the first years of operation, the farm distributed 600 eggs and 580 pheasants. Within a year, the volume grew to 1,220 eggs and 1,559 birds shipped to various district wardens, who released them when possible on protected farms.

The commission extolled the farm's progress in the 1926–28 biennium, saying it "made a good showing" despite unusually heavy and extended rainfall with occasional heavy hail that killed several hundred birds: "A fair production was had and each county received its quota of some 35 birds" from 2,230 pheasants and 1,424 eggs in 1926-27 and 2,213 birds and 500 eggs in 1927–28. Even so, the commission felt compelled to defend its pheasant program, saying in its seventh biennial report,

> Since the department started the introduction of ringneck pheasants four years ago, there has been much speculation as to whether this species could ever be successfully acclimated to our covers. At this time the undertaking is largely experimental and it will doubtless be several years before results can be accurately predicted. Reports during the past two years have been much more favorable than those of the first half of the period.
>
> Very encouraging reports are coming in from those sections where reasonable care was used in the selection of covers and where proper protection from hunters, fires and predatory animals has been given.

Pheasant

A ring-necked pheasant at the State Game Farm near Rogers, 1925. During the eight years the game farm was operated, 18,192 pheasant were reared and released around the state. Attempts to permanently establish the bird in Arkansas were unsuccessful.

Arkansas Game and Fish Commission.

With our vast acres of wild land and abundance of predatory species and a large number of careless hunters who bang away at a pheasant or anything else just to see if they can hit it, the introduction of a new species is more than a gigantic task. It is an uphill fight similar in many respects to that being waged here . . . by a few little bunches of prairie chickens that are struggling for existence and have not made any material progress in 10 or more years.

The commission slowly was learning the value of habitat, as illustrated in its report covering the 1928–30 biennium:

> The State Game Farm experienced two very successful seasons, as a result of which the covers of Arkansas received a total of 4,625 ring-neck pheasants.
>
> A different plan for distributing birds was followed during the past two years. Some six or eight suitable sections of the state were selected and all of the birds produced from the Game Farm were planted in those sections. It seems reasonable that a substantial number of birds planted in a limited number of areas would, in the long run, show better results than a very limited number . . . planted in each county of the state. As the birds become somewhat established in the areas selected, other sections will be given attention.

The commission said it was encouraged about its pheasant program. "From a number of sections come reports that the birds are showing agreeable progress. These reports come largely from areas that are not hunted at all, from which predators are regularly killed and which are highly cultivated for agricultural or horticultural purposes."

When the Depression hit, however, the commission no longer could justify the game farm's expense, and its operation ended in 1932. The legislature authorized its sale in 1939. In the last biennium before the closing, 2,442 pheasants were distributed in 1930–31 and 2,450 in 1931–32.

MOURNING DOVE

This migratory bird was so plentiful in Arkansas its population warranted little or no concern until a sharp decline was noted in the early 1950s. These fast-flying birds are hunted principally in the Delta, ideal habitat for the seed-eating dove (Holder 1951).

RUFFED GROUSE

No known records indicate the existence of ruffed grouse in Arkansas after 1900. However, old hunters told of killing up to thirty birds in a single day up to that time, which probably helps explain their disappearance. Yet this cannot be the entire story because ruffed grouse are extremely wary and take cover in dense woods in rough terrain. Destruction of timber, burning, vegetation changes and intensive grazing may be more plausible explanations. No effort was made to restock the ruffed grouse in the 1915–45 period.

BIG GAME

BLACK BEAR

The Arkansas Game and Fish Commission reported in 1928 that there were about twenty-five bears in the state. None had been seen except in the lower White River country where they ranged on both sides as far north as St. Charles, although "considerable bear sign" was found by a state trapper in the hill country west of Hot Springs.

Black bears, the commission said in its seventh biennial report, were protected fully by law after 1931; however, sentiment favorable to protection had not crystallized as rapidly among Arkansans as legislative action indicated:

> As a general rule the average person believes in killing a bear on sight and, if one is not easily sighted, there is strong inclination to get the dogs and chase him down. When brought into court for ending the life of a bear the killer usually proves that the animal was destroying crops, poultry or livestock for him and thus creates a knotty problem for the judge or jury. We are hopeful that as the black bear is re-established in considerable numbers he will grow in popularity and his protection become assured from the public sentiment standpoint.

BISON (BUFFALO)

The bison, or buffalo, had been killed or driven from Arkansas by the mid-1800s. The Game and Fish Commission, without explanation of its expectations, obtained one buffalo through the U.S. Bureau of Biological Survey in 1929 and placed it at Lake Catherine in Garland County.

ELK

The commission obtained an elk from the Bureau of Biological Survey in 1929 and placed it at Lake Catherine.

Elk release

A cow elk from Oklahoma awaits release on Black Mountain Refuge in Franklin County in 1933. Such releases were part of a U.S. Forest Service restoration effort.

Arkansas Game and Fish Commission.

The U.S. Forest Service attempted to restore elk in the Ozark National Forest in 1933. Three bulls and eight cows were brought from the Wichita National Game Refuge near Lawton, Oklahoma, and placed in the Black Mountain Refuge in Franklin County. A special study in 1943 revealed the presence of about seventy-five elk in Arkansas, all of them in and around the Franklin County refuge, with occasional wanderings into Washington and Madison Counties (Holder 1951). One cow elk was reported in Johnson County in 1943.

WHITE-TAILED DEER

The comeback of deer in Arkansas is one of the state's most dramatic success stories. The Game and Fish Commission did not know specifically how many deer were in the state in 1915, only that there were few, and the demand for them by sportsmen was high. Therefore, the agency turned its attention to deer long before it did to wild turkeys and other game. There was no sense in restocking, however, until laws were tightened. The 1917 legislature was persuaded to close the season on does and limit the season on bucks until 1922 to protect the estimated two thousand deer in the state (Holder 1951). The two-deer bag limit was extended, but the legislature failed to prohibit chasing deer with hounds (as it had done earlier in several counties or parts of counties) as the commission requested.

Nothing seemed to work. By 1925, the commission began advertising through its magazine, the *Arkansas Conservationist*, and other publications its intent "to purchase every available animal for brood stock." If there was a surplus, the commission promised the deer "would be liberated in protected territory." Persons having any deer at all for sale were urged to call. Arkansas was not the only state desperate to find deer. The commission received a request for a permit to ship several deer to Pennsylvania. It was a violation to ship any game out of state without permission. Not only did the commission deny Pennsylvania's request, but it bought the deer "for propagation purposes."

It was hard to procure deer and turkeys the commission admitted, and the ones it did get came "at a comparatively high cost." Deer were bought and shipped in from Minnesota, Missouri, and Louisiana in 1926–27, and, through the cooperation of the U.S. Forest Service, thirteen fawns arrived in Arkansas from the Pisgah National Forest in North Carolina. The Forest Service did the trapping, shipping, and stocking, while the commission paid the bills.

It was initially thought most of Arkansas's remaining deer herd was wiped out by the Great Flood of 1927. As the waters rose, deer scrambled atop levees where they were exposed to men who gathered to slaughter them. Sportsmen and commission wardens managed to save a few.

Even so, the deer herd dropped to its all-time low of about five hundred in 1930. Incongruously, the commission said in its 1930 report deer provided the "one ray of sunshine" through an otherwise "rather dark cloud." The deer comeback, the agency asserted, had made "wonderful progress" because there were reports from every section of the state that "deer were on the increase," and "we are encouraged to feel that the time is not far distant when Arkansas will, rightfully, take its place as one of the leading deer hunting states." Again in 1932, the commission reported deer were "gradually increasing in many sections." Two years later, the

Deer crates on trucks at Bald Knob

In February 1943, white-tailed deer purchased in Wisconsin were loaded in crates and shipped by rail to Bald Knob. Sixteen were carried by truck and mule wagon and released in the White County Game Refuge along Glaise Creek. Restoration efforts such as this helped deer stage a remarkable comeback in the state.

Arkansas Game and Fish Commission.

Deer crates being hauled by mule team

Deer release

commission was still saying deer were "showing an appreciable increase in many sections of the state."

As with other species, refuges seemed to be the turning point. In 1939, it was decided to carry the refuge idea a step further, and a deer farm was established at the Howard County Refuge. How farm deer were treated and what happened to them was described by Stella Wynn, a Dierks teacher, in the *Arkansas Historical Quarterly* (December 1943):

> Although there is a bountiful supply of natural feed for the deer, . . . small fields of corn are planted to supplement the natural feed. The idea is to have the brood stock in good condition at fawning time.
>
> Each winter along about Christmas or the first of the year, deer trapping begins. The surplus deer are taken out to parts of the states where no deer or very few deer exist. The trapping is usually completed by the last of February.
>
> The deer are shipped or hauled in dark crates. If the deer see light, they may injure themselves trying to get away. The doors of the crates fit the doors of the traps, and it becomes easy to slip the deer from the trap to the crate.
>
> At top production the farm plans to furnish as many as 50 deer per annum for planting in other states. This mark has not yet been reached, but will be soon.
>
> Taking the census each year is interesting. The game wardens and their supervisors hold a district meeting, a sort of wildlife institute . . . After the work is done there, the men are numbered and stationed equal distances apart across each division of the farm. The men then move slowly toward the opposite side of the enclosure. The deer also move in that direction, but when they reach the opposite side will turn back. It is then that the accurate count is made. Each man counts all the deer that pass to his right.

The commission also reported in 1932 it was beginning to win its campaign against killing does. "While there are many hunters who disregard the law which prohibits the killing of doe, many have come to look upon the killing of the mother or producer as unsportsmanlike." It called the attitude one that was "gradually increasing" and that would "aid materially in increasing the deer." The commission learned later the public could be oversold, and this could become a deer management problem itself, especially when conditions changed. This happened in Pennsylvania, Michigan, Wisconsin, and Minnesota where deer were never as abundant in frontier days as they became in the 1960s. Each state became overpopulated, with Pennsylvania and Michigan reaching such extremes as to be classic examples of mismanagement (Black 1968).

Does zealously were protected, and sportsmen in Pennsylvania learned to abhor killing any deer except legal bucks. Hunters had become overeducated, and deer in Pennsylvania became so plentiful near refuges it was not unusual to see thirty to fifty in one herd. Sometimes more than one hundred deer were seen at one time in summer or early fall.

When the Pennsylvania Game Commission proposed to open the entire state for one week before the regular season, during which does could be killed to reduce the herds, sportsman protested bitterly, and the agency was forced to take another more expensive, less efficient tack ("Pennsylvania's Deer Problem," 1927).

The Arkansas General Assembly banned using dogs to chase deer in White County and two Jackson County townships as early as 1901, well before there ever was a state Game and Fish Commission. The move proved so unpopular, however, the law was repealed two years later. In 1917, the commission asked the legislature to enact a law prohibiting using dogs to hunt deer. The recommendation got nowhere. Available evidence suggests those given the responsibility of protecting or managing Arkansas's wildlife have never wavered from the view that using dogs for hunting is detrimental to deer populations. This was the position in the early 1900s, and it was the view in 1951 when Trusten Holder wrote the following in *A Survey of Arkansas Game*:

> The Game Fish Commission in its second annual report asked sportsmen and hunters to wake up to the fact that game was disappearing and to make an organized effort to save what was left. The need for a large preserve was stressed, and again the Commission urged, "Eliminate the hound from the chase." In this and subsequent reports, the disastrous effect of dogs on deer was pointed out. This story was told to emphasize the situation: "There is one place covering an area of about four square miles in eastern Arkansas where a few deer are left. During the season of 1920 there were camped not less than 300 hunters on the spot, and in those camps they had about 200 hounds. There was one camp with 21 hounds and about the same number of deer. Deer in the mountains were continuously chased by dogs. There were no large lakes for the deer to swim into and elude their pursuers, so they fell easy victims."

This position has just as consistently been met and defeated by vociferous proponents of one of Arkansas's deepest traditions, particularly in mountainous areas.

The Game and Fish Commission's assistant chief of wildlife management, Jim Spencer, writing in *Arkansas Game and Fish* (September/October 1984), said the problem is not as great in the flatlands because of the terrain:

> In the mountains, deer have regular crossings that have been used for centuries, and it's a relatively simple matter for a hunter or group of hunters to cut off the escape route for deer running along a particular ridge or hollow . . . By contrast, deer in the heavily wooded flat areas . . ., such as the Gulf Coastal Plain in southwest and south-central Arkansas and the Delta country of the eastern part of the state, have much less predictable travel routes. They have much more cover for escape, and have an unlimited choice of escape routes, including many creeks, lakes, swamps and bayous which often enable them to lose a pack of trailing dogs.

Dog deer hunters counter that the "experts" do not have enough valid evidence to support their antidog views, and a few have gone so far as to say they would rather have no deer at all than have the government dictating their method of hunting.

SMALL GAME

SQUIRRELS

Arkansas had a "considerable supply" of gray and fox squirrels in 1915 even though the law allowed them to be sold within the state, causing them to be slaughtered in "great quantities" for the market, the commission reported. It noted that

Squirrel hunter draped in squirrels

Prior to 1917, there was no bag limit on Arkansas squirrels. Hunters killed them in great numbers, and many were shipped to game markets in the state's larger cities.

Arkansas Game and Fish Commission.

during the previous two springs, squirrels were shipped at a rate of about five hundred a day from an unnamed town in Arkansas to markets in the state's larger cities.

The law allowing intrastate squirrel sales was not repealed until 1927, but the commission got the General Assembly to impose a bag limit of fifteen in 1917. However, before White River National Refuge was established in 1935, it was not uncommon for a hunter to kill more than fifty squirrels in one day in that area. The bag limit was reduced to eight in 1927, the same year the Delta flooded and large numbers of squirrels were killed by hunters in boats. As a result, the legislature also made it unlawful to shoot squirrels from boats. This regulation has been changed since to allow hunting from boats so long as no motors are used while hunting.

Squirrel populations are cyclical. Migrations appear to be the direct result of overabundance at the peak of population cycles. At such times, large numbers of squirrels start across country, all moving in the same direction and proceeding despite any and all obstacles. These movements still occur, but in only a minor way compared with the great migrations of the past (Black 1968).

A substantial mass movement of squirrels occurred in 1936 in Howard, Pike, Polk, and Sevier Counties. They were said to have come either from the north or northwest and stayed for about two months. Even after the main movement passed, squirrels remained abundant in those counties. Perry County also reported a squirrel migration in 1936, but it was not judged to be part of the same movement that occurred farther southwest. A gray squirrel migration apparently took place in the Ouachita River Bottoms in south Arkansas in 1942 because a refuge keeper there reported seeing as many as three hundred squirrels a day. The mast crop was poor that year, and the squirrels disappeared but were returning to their normal density by 1943 (Black 1968).

RABBITS

Like squirrels, Arkansas's supply of rabbits was "considerable" in 1915, but the Game and Fish Commission clearly was alarmed about the magnitude of the cottontails' slaughter for markets in and out of the state. Twenty-three states had laws regulating shooting of rabbits, but Arkansas was not among them.

The commission returned to the subject of rabbits in 1928, noting again the cottontail had never been placed "under the protective wing of the law." Even though there was no great demand for protecting rabbits, the commission predicted the day would "undoubtedly come when this now disregarded rodent will be in popular demand by our hunters."

FURBEARERS

Furbearers were important during the exploration and settlement of Arkansas but received scant attention after their skins ceased to be a principal medium of exchange and provide a livelihood for a significant segment of society. As the country became dominated by agriculture, industry, and urban centers, furbearers received little appreciable notice except when trappers and deer hunters or other houndsmen clashed.

This is not to suggest furbearers were no longer important; perhaps they just did not receive the attention they once did. A trapline requires hours of plain hard work. Some did it for recreation and the opportunity to be outdoors. For many, the extra income from a trapline was often a buffer against hardship or the means of paying for an education. Others trapped out of necessity, to protect crops and livestock against raids by everything from raccoons to rats.

The Game and Fish Commission estimated in the mid-1920s Arkansas contributed about $500,000 in furs to the world market. By 1928, the commission was saying it was "conservatively" estimated that Arkansans received $3 million from pelt sales during the 1927–28 season. As it had before, the commission said a lot more would be realized if Arkansans would "put forth a special effort to protect pelt-producing animals." "Unfortunately," the commission continued, "many good people occasionally go out early in the fall, before the legal open season, and take furbearers before the pelts have become of any value whatever. It may be safely said that the same coon pelt that sells for $1 in October would bring $6 to $8 if left on the animal until the middle of December."

The value of Arkansas furs in the 1950s was estimated to be about $1 million annually (Holder 1951).

RACCOON

In the 1920s raccoons usually ranked first or second in the number of pelts sold, in part because they range statewide and never approached extinction. The lowest population point occurred about 1935 during the height of the Depression.

In 1885, a raccoon pelt brought $.25, a good price in those days. In the 1940s, when World War II raised both prices and the number of pelts taken, average prices ranged from a high of $3.72 in 1943-44 to a low of .$53 in 1949–50 (Holder 1951).

Raccoon pelts at a fur storage facility, 1950

In 1950, the average price for a raccoon pelt was fifty-three cents. Although raccoons have never been scarce in Arkansas, in 1942 the Game and Fish Commission obtained federal aid to trap 250 raccoons for release in game refuges around the state in an attempt to "increase the breeding stock."

Arkansas Game and Fish Commission.

BEAVER

Ernest Thompson Seton (1929) in *Lives of Game Animals* estimated there were no fewer than 60 million, and possibly as many as 400 million, beavers in North America before settlement. They were trapped heavily and almost had been wiped out by the early 1900s. The Game and Fish Commission reported in 1915 the state had fewer beavers than it had bears. The commission announced with obvious pride in its 1926 report that, through the courtesy of D. H. Madsen of Utah, it had obtained four beavers and released them in Big Lake. "Beaver formerly inhabited Arkansas in great numbers and it is hoped to bring them back in suitable areas," the

Beaver release at Snake Creek

Beavers shipped from Pennsylvania are released in Howard County's Snake Creek in April 1932. Such early restoration efforts were unsuccessful. The Game and Fish Commission reported that there were only sixty to seventy beavers in the state in 1946.

Arkansas Game and Fish Commission.

agency said. The Big Lake beavers soon vanished. A serious restocking effort did not start until years later, but its success, aided in large part by encroachment of beavers from Mississippi and Louisiana, probably dwarfs that of any other restoration undertaken by the wildlife agency.

OPOSSUM

Too common to attract much attention, the opossum usually ranked first in numbers trapped among furbearers and occasionally ranked first or second in value. The city of Mena (Polk County) recognized the value of the opossum when it held an annual festival with an elaborate banquet in the animal's honor.

The price of opossum hides in the 1940s ranged from low of nineteen cents in 1940–41 to a high of sixty cents in 1945–46 (Holder 1951).

MINK

A trapper reported in 1930 he caught fifty mink, and in 1945, when he worked just as hard, he caught only four. The low point in Arkansas's mink population apparently occurred about 1935, again during the height of the Depression. The Ozark and Delta regions produced the largest mink catches.

Prices for mink pelts have varied widely, but they generally are the most valuable of any Arkansas furbearer. Prices went from a low of $5.11 in 1940–41 to $19.64 in 1948–49 and began a slow decline from that point (Holder 1951).

SKUNK

For reasons that need no explanation, skunks were trapped less than other furbearers until about 1919 when increased pelt prices made them worth the trouble. In the 1940s, skunks ranked fourth in the number of furbearers trapped, and pelt prices ranged from a low of $.45 in 1949–50 to a high of $1.45 in 1943–44 (Holder 1951).

MUSKRAT

A Survey of Arkansas Game (1951) said there seemed to be a natural boundary within Arkansas for the prolific muskrat because its southern limit was Lee County. No native muskrats were known to exist in the Gulf Coastal Plain. The author of the work speculated that the Arkansas River and the Ouachita Mountains might have acted as a natural barrier that was by-passed when the Game and Fish Commission released some in Petit Jean State Park in Conway County and Bragg Lake in Ouachita County, where they established themselves. Muskrats were abundant in Big Lake National Wildlife Refuge, and animals stocked south of the Arkansas River in 1942 and 1943 came from this area.

Prices for muskrats ranged from a low of $.61 in 1940–41 to a high of $2.10 in 1947–48.

RIVER OTTER

The river otter was trapped extensively by Arkansas's Quapaw Indians and by white trappers who explored and helped settle the state. Like beavers, otters were almost extinct by 1900, though one trapper reported catching fifteen in 1915 (Holder 1951), the same year the animal was first legally protected in Arkansas. Prices for otter pelts ranged from five to fifteen dollars (Black 1968).

Rigid protection, coupled with new habitat created by beavers, has started the otter on its way back in parts of the United States, including Arkansas. An estimated seven to eight hundred otters existed in Arkansas in the late 1940s, mostly in the Delta on the Cache, St. Francis, White, and other rivers. The densest concentration was reported at that time in Ouachita County, where they ranged along the Ouachita River, Tulip Creek, and Bayou Freeo. An otter was reported killed on a small tributary of Spring River in Sharp County in January 1946 (Holder 1951).

FOXES

Foxes are furbearers and predators, hence their location here, bridging the two classes.

The red and gray fox market collapsed in the late 1940s and early 1950s, skidding to a low of \$.43 in 1949–50 from \$6.47 in 1943–44 for the red fox and to \$.25 from \$2.09 for the gray (Holder 1951). The falling prices helped fox populations recover somewhat in Arkansas.

Red and gray foxes prey largely on rabbits, mice, other rodents, and insects. During the 1934–36 drought in Arkansas, the fox, then under full protection in the state, reached peak abundance, while rodents virtually disappeared. For a year or two, foxes preyed heavily on quail and were "unquestionably instrumental in reducing their numbers." This situation was "decidedly abnormal, and there is no justification for the generalization that foxes are responsible for low quail populations" (Black 1968). Overshooting and especially habitat loss are what experts saw as the reason for quail shortages, more so than fox predation.

Fox hunter

Numbers of foxes showed remarkable increases following several years of scarcity after World War I. The unidentified man in this photo wrote to the Game and Fish Commission to complain about poultry depredation. "The foxes pictured . . . in three weeks caught 38 of my chickens valued at \$57. After killing these three, I lost no more chickens. Had they [the foxes] been prime, they would have been worth about \$3."

Arkansas Game and Fish Commission.

Red foxes were not distributed as widely nor were they as numerous in Arkansas as gray foxes. Foxes became scarce in Arkansas after World War I, and the Game and Fish Commission became interested in protecting them in the 1920s. Both species increased "remarkably" during the 1930s and 1940s (Holder 1951).

Arkansans still trap gray foxes, and there is a deep-seated tradition of fox chasing as described in part 1 by J. J. Propps. Many Arkansans thrill to the sound of dogs chasing a fox. The crisp night air, the smell of fresh coffee, the fellowship, and the barking of dogs make fox hunting a popular sport.

Not long after World War II, houndsmen had to become content with chasing the gray fox instead of the

Dead bobcats

more popular red, which went into a decline for reasons that have defied firm explanation. One possible explanation is the eastward expansion of the coyote, which occupies similar habitat, is larger, and competes better for food and den sites (Pledger 1975).

Kill a bobcat, save a turkey. Kill a wolf, save a pig. Such was the attitude of many Arkansans who fostered a shoot-on-sight mentality that led to the wholesale slaughter of predators in the state. This attitude prevailed until the 1970s when the public gained a greater understanding of the predators' role in maintaining ecological balance. By then, the red wolf was extinct in Arkansas.

Arkansas Game and Fish Commission.

PREDATORS

BOBCAT

While it no longer commands much public attention, the bobcat was mentioned frequently as a major constituent of Arkansas wildlife in the state's early history as citations in part 1 illustrate. Predators, including the bobcat, were hunted and killed

Wolf hunters with pack of hounds, dead red wolves

relentlessly, and those who brought them in were praised by their communities. For example, the *Helena World* newspaper reported on March 16, 1898 that

> A negro named Walker came to town about noon today with the body of a wildcat he had killed about two miles west of the city between Mr. H. P. Grant's and Mr. M. J. Crocker's places. Walker says he killed it just after daybreak this morning. The animal was discovered by Walker's dogs, which had chased it up a tree, where the negro shot it . . . It was a female of pretty good size and, according to Walker, the male wild-cat which is in the same neighborhood, and has been seen by several persons, is considerably larger.

This attitude prevailed until the public gained greater understanding of the predator's role in maintaining environmental balance. The bobcat is a kind of biological governor that keeps in check populations of small, prolific breeders such as rabbits and rats (Dean 1985).

In the mid-twentieth century, bobcats were reported to be fairly common in isolated areas, and areas less remote had a few that ranged in and out of large timbered tracts. They were not protected by law (Holder 1951).

PANTHER

In part 1, this animal is referred to variously as the mountain lion, cougar, painter, and puma.

Though one of the shyest of animals, the panther has been one of the most feared and relentlessly exterminated animals in American history. It once was thought to be extinct in all regions east of the Rocky Mountains except for one or two small areas in Florida. Stories of panthers still persist throughout Arkansas (Black 1968). Panthers, like bobcats, can live almost undetected in an area for years.

The game survey done in Arkansas and published in 1951 stated "the last of the native panthers seem to have disappeared in the 1920s," but that two were taken in the Ouachita Mountains in the late 1940s. Another, now on display at Game and Fish Commission headquarters, was killed near New Blaine in 1976 (Holder 1951).

RED WOLF

Red wolves have been exterminated but were said to have lived in the rugged terrain of southern, northern, and western Arkansas with the largest populations in the 1940s in Benton, Carroll, Crawford, Fulton, Johnson, Sharp and Washington Counties. These areas had ideal wolf habitat—small farms scattered in large wooded areas. Sharp County probably had more wolves than any other, with estimates ranging from forty to one hundred. In 1943 and 1944, thirty-two wolves were trapped in Washington County when the Game and Fish Commission responded to complaints from farmers and livestock raisers.

Wolves also were reported during this time in Calhoun, Columbia, Grant, Hempstead, Lafayette, Little River, Miller, Ouachita, and Sevier Counties in the Gulf Coastal Plain. A few wolves were reported during the 1940s in Howard County in the Ouachitas, and nine were trapped in Pulaski County in 1942–43 (Holder 1951).

COYOTE

The coyote, an emigrant from the west, began extending its range into Arkansas in the mid-twentieth century with the cutting of hardwood forests and creation of open lands. Large specimens seldom exceed thirty-five pounds. A large male taken in Franklin County in 1951 weighed thirty-four pounds.

The current status of both the red wolf and coyote will be discussed in part 4.

WATERFOWL

DUCKS

Throughout this century, Arkansas has been known as one of the finest duck-hunting states in the country. The reputation is deserved.

In the 1700s, a French explorer complained that ducks were so thick on the Arkansas River he could not stir the water with his paddle. Two hundred years later, market hunters were taking as many as twenty-five thousand ducks a day from Big Lake in Mississippi County for meat markets in St. Louis, Memphis, New Orleans, and Chicago (Fleming 1972).

When bag limits were first set under international treaty, a daily limit of twenty-five ducks with a possession limit of fifty was allowed. In more recent times, the limit has never been more than seven, except during World War II when a meat shortage prompted the federal government to raise it.

Several duck hunting clubs existed in western Arkansas about 1900, but most were disbanded by the 1940s. About 90 percent of the duck hunting in Arkansas is done in the twenty-one counties of the Delta. Within the Delta, the Grand Prairie area around Stuttgart, DeWitt, and Gillett has given the state its international fame for duck hunting.

Rice shocks near Stuttgart

Before modern combines were invented, rice grown on the Grand Prairie was cut and put into shocks. Ducks flocked to the shocks to feast on the rice, and hunters flocked to the rice fields to shoot the ducks. It was during this period that Stuttgart gained a reputation as the mallard hunting capital of the world.

Arkansas History Commission.

Before modern, efficient combines were developed to harvest crops, rice grown on the Grand Prairie was cut and put in shocks. A field would have hundreds of shocks. "They let the rice sun-dry, which was just like setting the table for the ducks," W. M. Apple of Little Rock explained in 1972. "The ducks would feed on rice all night long and then at daybreak they would head for the White River bottoms. Then they fed on acorns during the day. This was where most of the hunting was done. The [hunting] guides would have boats on the river and they would take their parties down the river to these old river lakes" (Fleming 1972).

In 1925, rice farmer Arthur Tindall conceived the idea of impounding rain water in reservoirs so he would not have to pump so much groundwater to irrigate his crops. Though he was not thinking of ducks when he did this, the levees built on then-cheap timberland formed lakes, and the ducks flocked in. From the ducks' standpoint, it was a cozy situation because they could gorge on Tindall's rice and then rest in the cover of his flooded timber. Frank Freudenberg followed Tindall's example in 1933. He built the first of several artificial lakes on his timberland, and more ducks came and stayed (Coghlan 1949). There are more than one hundred of these reservoirs in the Stuttgart area now. Every fall, ducks flying south from Canada in the Mississippi Flyway stop to fill up on rice and acorns. In later years, some lakes were built to attract ducks, not to furnish irrigation water. These are called greentree reservoirs, where water is pumped in before duck season and released after it is over. Thus it does not kill the trees, and they remain green.

Even without rice and reservoirs, ducks would be plentiful in eastern Arkansas: "It is more than the coincidence of adjacent marshes and rice fields that makes the region around Stuttgart . . . the world's finest duck-hunting grounds. Stuttgart is in the heart of the Mississippi flyway at what is perhaps its narrowest place, along a route followed by more migratory birds than any other in the world," Dr. John D. Black wrote in *The Management and Conservation of Biological Resources* (1968).

Because many of the best hunting areas are privately owned, duck shooting has been said to be a rich man's sport. An article in the *Arkansas Gazette* on November 26, 1935, attests to this. It reported that

> Herbert Pulitzer of New York and Joseph Pulitzer of St. Louis, sons of the famous New York publisher, have gone in for duck hunting in a big way.
>
> They have rented the ground floor of the Riceland Hotel [in Stuttgart], also two houses in Stuttgart and leased a 1,500-acre tract, including a large reservoir on the rice plantation of Frank Freudenberg, six miles east of Stuttgart.
>
> It is reported that the brothers have installed a retinue of attendants, including a hairdresser, in the hotel while they and their wives and guests are occupying the homes that they have rented.

An undated newspaper clipping, "Joe Pulitzer Leases Freudenberg Reservoir," found in the Freudenberg scrapbook and made available by Bennie Burkett, concluded that, "Off-hand, I'd estimate he [Joseph Pulitzer] spent a good many thousand dollars of benefit to the State Game Department and the town and community." "Pulitzer quit shooting every day at 10:30 a.m., rested his place regularly, and I doubt if his total bag (much of which went to hospitals and poor people) exceeded 1,500 ducks, probably less."

What was meant by Pulitzer "resting his place regularly" was explained by Ralph Coghlan in the *St. Louis Post-Dispatch* on December 30, 1949:

> In most duck clubs, great care is taken not to overshoot the area or, as the phrase goes, to "burn it out." Repeated shooting at any one place where ducks gather causes the ducks to leave and not to come back. Therefore, these areas are rested. Moreover, much of the shooting at Stuttgart is very sporting, and there are many more misses than kills. To knock down a wild duck in full flight over the tops of tall trees, while the hunter himself may be standing in water up to or over his knees, is not the easiest thing to do, as this city boy can testify.

It was estimated in the 1970s that equipment alone cost a minimum of $200.00. Using a guide could cost $17.50 to $20.00 a day then, and the going rate is more than twice that today. "The ducks you shoot will cost anywhere between $5.00 and $50.00 a pound depending on how much you hunt during the season. However, "it's the sport and not the meat that really counts," the *Gazette*'s John Fleming commented.

Duck hunting was a sport that even captured the archer. A newspaper clipping from an unspecified publication, "New Precedent for Duck Hunters," dated November 28, 1940, found in the Freudenberg scrapbook, reported, "Larry Whiffen, crack shot with bow and arrow, was the first person to ever kill a duck with an arrow. This feat was performed at the Freudenberg Reservoir near Stuttgart on Nov. 15 [1940], and was a feature of the World Championship Duck Calling Contest."

The attention paid to Canadian habitat and the relatively early institution of bag limits under federal law apparently paid off. A 1971 Little Rock newspaper article ("Duck Flight Near Record," reprinted in the *Stuttgart Standard,* December 16, 1971) reported that "the greatest flight of ducks into the rice region since the record year of 1917 was reported this morning from the lake and reservoir regions." The report continued:

> Countless thousands of the fowls have been here for several days and the cold weather in the North brought an additional flight yesterday. Harvey Tow, deputy sheriff in charge of game violations at the large reservoir of Frank Freudenberg, four miles east of Stuttgart, reported this morning that yesterday's flight practically doubled the number of ducks which for several days had been near normal proportions.

Ralph Coghlan, editorial page editor of the *St. Louis Post-Dispatch,* in a column appearing on December 21, 1947, was sardonic about the mass of Stuttgart ducks:

> let me say that, over a hunting experience of many years, I have never seen more ducks than darkened the Arkansas skies this year. Not being a bookkeeper, an accountant, a human adding machine or a member of the federal Fish and Wildlife Service, I couldn't come within 100,000 of figuring how many I saw.
>
> I remember that last January the Fish and Wildlife Service took a dim view of waterfowl prospects, estimating that the waterfowl population had declined 26,000,000 from last year's estimate. Then, in August, the service reported that nesting conditions in Canada had been found "extremely discouraging," and Albert M. Day, director of the Service, said that "we are in for a poor shooting season." . . . As a result, the bag limit was reduced to four ducks a day, with a possession limit of eight.
>
> It may be as some say, that the floods of last year caused the ducks to scatter widely, while this year's dry weather caused them to concentrate more thickly. Anyway, I can cheerfully report that the magnificent mallard is in no danger of extinction. I watched mallards sitting in vast and solid rafts on the Arkansas reservoirs quacking raucously and happily, and at dusk, saw them start for the rice

> fields. They took off in successive roars like fleets of miniature B-29's, and for half an hour or more the whole sky was alive with ducks. Whether you watched them from Al Meehan's picture window overlooking the Tindall place, or at Frank Freudenberg's or Peckerwood, or Lost Island, the sight was the same. And, then at dawn, as you paddled through the timber, wave after wave of mallards got up like thundering aerial herds. In other words, there are several ducks in Arkansas.

This was written seventeen years after the 1930 drought all but gave the *coup de grâce* to ducks, thoroughly arousing sportsmen and conservationists to support passage of the Duck Stamp Act.

CANADA GEESE

Canada geese use the same flyways as ducks. The largest concentrations are along the Mississippi River from southern Illinois and Missouri southward into Louisiana (Black 1968). Because of a change in distribution patterns, however, the Canada goose population in eastern Arkansas dwindled, dropping to fewer than nine thousand in the late 1970s (Walter 1982).

MUSSELS

In early 1926, Thomas K. Chamberlain of the U.S. Bureau of Fisheries made a survey of Arkansas's shell-taking industry. Based on his findings, certain waters were closed to shell taking, effective January 1, 1927. The closure was in effect only a few months when the "disastrous spring floods" that year "so thoroughly demoralized conditions" along the rivers "as to render necessary an opening of . . . all closed waters."

High waters continued during 1927 and early 1928, and little shelling could be done. The Game and Fish Commission said it would take steps toward "proper preservation of the mussel industry" when the rivers and their drainage basins returned to normal. It did not happen.

The history of waste of this resource follows a familiar pattern: pollution, waste, overexploitation, and then complaints that the animals have become scarce. Kentucky, Arkansas, Indiana, Michigan, and Iowa were all once great producers of mussel shells used to make buttons. By the 1960s, there were only remnants of this once-flourishing trade, with the only button industry of any significance being in Tennessee (Black 1968).

FISH

Construction of the Lonoke Fish Hatchery and other early efforts to restock Arkansas's waters have been described in this part; more extensive treatment is given fish in part 4.

PART 3

Recovery and Conservation

11 Passage of Amendment 35: A New Game and Fish Commission

Amendment 35 was the hottest issue on the ballot in the 1944 general election. However, the Amendment 35 in the election had nothing to do with the Game and Fish Commission; rather, it was the question of whether Arkansas should adopt a freedom-to-work (known generally today as right-to-work) law by prohibiting closed union shops. The amendment that would revamp the Game and Fish Commission went before the voters as Amendment 36. When one of the proposed amendments on the ballot failed to pass, Amendments 35 and 36 were forced back a number, so the Game and Fish Commission came into being and continues to operate under Amendment 35. Amendment 36 was sponsored by the Arkansas Wildlife Federation, whose members had grown weary of seeing the state's wildlife resources sacrificed to politics—that is, the wholesale replacement of both commissioners and Game and Fish Department employees whenever a new person became governor.

Based on news coverage in 1944, Amendment 36 generated little controversy. The *Arkansas Gazette* (November 5, 1944) endorsed the amendment, and in a page-one box that listed ballot issues and the newspaper's positions, it said Amendment 36 was

> Designed to stabilize the Fish and Game Commission against shake-ups following changes in statehouse administrations and put fish and game conservation on the most efficient and practical basis. Vote FOR.

In an October 31, 1944, editorial, the *Gazette* said criticism of proposed Amendment 36 appeared directed primarily at the provision giving the Game and Fish Commission power "to make regulations that might vary in different parts of the state in different years." The newspaper insisted this was a good provision, not a poor one because flexibility was needed, "indeed is imperative, if regulations are to fit wildlife facts." The editorial continued:

> Game and fish regulations contained in laws passed by the legislature, which ordinarily can not be revised to meet changing conditions oftener than once in two years, may work against the very purpose of such laws—the protection of game birds, animals or fish.
>
> A year ago last spring, flood waters stayed so long that they interfered with the nesting of wild turkeys. The practical remedy for meeting this loss of the normal addition to the numbers of Arkansas' most prized game bird would have been to put wild turkeys on the protection list this year. But again the Game and Fish Commission was without power to act.
>
> These are . . . examples of unforeseeable conditions that may develop between legislative sessions, and may affect wildlife in the state as a whole, or in certain sections, without flexibility impossible under present laws there is no practical way of dealing with them.

The editorial pointed out that the U.S. Fish and Wildlife Service had flexible regulations, so there was no reason the state should not have the same.

Proposed Amendment 36 went to the secretary of state in July 1944 for certification of petition signatures to get the measure on the ballot. Constitutional amendments, initiated by the people, needed at least 9,888 signatures. Amendment 36 was submitted with 11,074 signatures gathered from fifty-four of Arkansas's seventy-five counties, 2,433 of them from Pulaski County.

In its brief description of the amendment printed on July 7, 1944, the *Gazette* said it was a proposal to "stagger the terms of the Game and Fish Commission and give the agency power to regulate the taking of game, fish and furbearing animals." Commissioners could be removed "only for cause after a hearing," the report continued, and the commission would have "the exclusive power and authority to issue licenses and permits, to regulate bag limits and the manner of taking game and fish and furbearing animals and shall have the authority to divide the state into zones and regulate seasons" and fix penalties for violations.

The first commissioners were appointed by the governor for staggered terms of one to seven years. One new commissioner is appointed annually to serve a seven-year term. The eighth commission member is the chair of the Department of Zoology (later the Department of Biological Sciences) at the University of Arkansas, who serves as an adviser with no voting power.

The Game and Fish Commission was charged broadly with the duty of protecting the state's wildlife and fisheries resources, setting and enforcing regulations and laws; operating fish hatcheries and game refuges; enhancing game propagation and restocking; managing fire protection; controlling stream pollution; and improving water quality (later transferred by statutory law to the Department of Pollution Control and Ecology); and controlling predatory animals.

The commission's chief executive officer was called the executive secretary until the title was changed to director in the early 1940s (though Amendment 35 has not been amended to reflect this). The director is appointed by and serves at the pleasure of the commission. The person in this post is responsible for executing the policies and regulations adopted by the commission.

The Game and Fish Commission operated almost exclusively from funds derived from the sale of hunting and fishing licenses, permits, and fines assessed for violation of game and fish laws. No money was received from the general revenue fund, and the commission was its own collecting and disbursing agent for these funds.

Just a year before the voters adopted Amendment 35 in 1944, they bestowed some insulation against politics on the boards of trustees of Arkansas's state-supported colleges and universities through Amendment 33, which was viewed as a reaction to Gov. Homer Adkins's stacking of the University of Arkansas Board of Trustees with appointees who would fire J. William Fulbright as University of Arkansas president because the Fayetteville newspaper his mother owned supported Adkins's Democratic opponent, former governor Carl Bailey. Eight years after Amendment 35 was adopted, Arkansas voters accepted the so-called Mack-Blackwell Amendment that made the Highway Commission an "independent" agency—that is, one "removed from politics."

While it is true the Game and Fish Commission has its own sources of revenue and acts as its own disbursing agent, it clearly remains a state government agency that can spend nothing unless an appropriation law of the General Assembly is passed. This keeps the commission responsive to the legislature when issues become controversial. In addition, there can be no increase in charges for resident hunting and fishing licenses without legislative approval. This, too, has been a legislative bargaining tool and check on the commission. An example of legislative control appeared in this lead article in the *Arkansas Gazette* on February 25, 1953:

> The Arkansas Game and Fish Commission's biennial appropriation was beaten for a second time yesterday in the House of Representatives, setting off a chain reaction threatening all remaining appropriation bills.
>
> By midafternoon there was reaction to what appeared to be a signal victory for a hard-core of representatives critical of the Game and Fish Commission and its administration.
>
> Using the same tactics employed to beat the conservation agency's bill, another block either voted negatively or abstained from voting in sufficient numbers to beat the proposed $100,000 appropriation . . . for the governor's emergency fund.
>
> Opposition to the [Game and Fish] measure was spearheaded by Representative Paul Van Dalsem of Perry County, perennial critic of the Game and Fish Commission, and Representative G. C. ("Jack") Carter of Franklin County.
>
> Van Dalsem promised the House that if it would beat SB208 that another appropriation bill would be introduced in the amount of $1,391,000 "to take care of your county game wardens and to provide more money for building lakes and hunting grounds." Carter voiced the same promise and a bill (HB558) was introduced in that amount later in the day.
>
> Meanwhile, the Game and Fish Commission—meeting in extraordinary session—announced that a new budget bill for the same amount and purposes, with one exception, would be introduced in the legislature in the next day or two. The one exception is a $6,000 a year job. It would be a public relations officer and assistant executive secretary to aid Executive Secretary T. A. McAmis, who has been one of the principal targets of disgruntled legislators.

A review of *Arkansas Gazette* issues during the early 1950s revealed it was not unusual for the General Assembly to balk at approving the Game and Fish

Commission's appropriation bill, and frequently, the action was a result of legislative unhappiness with McAmis. The newspaper seldom reported why the legislators were dissatisfied with McAmis.

As to the legislature's power to set resident hunting and fishing license fees, commission chairman Hays Sullivan told reporters in 1984 he had made a unilateral covenant with some legislators to exempt given areas from restrictions against hunting deer with dogs in exchange for their support to raise the resident hunting and fishing license fees to $10.50 each. Sullivan said the commission had to have the additional revenues and that making such agreements was a way to achieve it.

The governor can and does affect policy through his appointments. This has been especially true in recent years when the commission was split three to three on some issues. Andrew Hulsey was fired as director in August 1979 when the commissioner appointed by the governor fulfilled a commitment to his supporters to become the fourth vote to oust the director.* Another 3 to 3 split occurred in 1983 when it was believed Sullivan only needed one more vote to reverse the commission's opposition to the plan by the Army Corps of Engineers and the Cache River-Bayou DeView Improvement District to channelize the Cache and its principal tributary. The reversal did not occur.

Politics cuts both ways, however. Riley N. Donoho of Fort Smith was appointed to the commission in 1956 by Gov. Orval E. Faubus. Asked during an interview in December 1985 how he obtained his seat on the commission, Donoho explained: In 1954, Faubus was opposing Gov. Francis Cherry in the Democratic runoff primary. The Huntsville newspaper publisher and former highway department director stopped in Fort Smith to meet with and solicit the support of the powerful federal judge John E. Miller and his fishing buddy, Donoho, a well-to-do and well-connected Fort Smith insurance man.

According to Donoho, Faubus asked Miller what he would want in return for his support in what had become a bitter runoff primary. Miller replied that he wanted only one thing, which was to have Donoho appointed to the Game and Fish Commission. Asked why that appointment was so important to Miller and to him, Donoho said it was because he was something of a zealot about getting more game and fish in the state. A description of some of his work in part 4 of this volume shows how serious Donoho was about this.

Evidence suggests T. Arthur McAmis, in his capacity as director, was responsible in part for projecting an aura of independence in the new commission's first decade. An Alabama native who had lived in Arkansas since the late 1920s, McAmis joined the Game and Fish Commission in 1941 as an auditor. Within a short time, he was appointed executive secretary in a staff reorganization. McAmis died unexpectedly of a heart attack in May 1956. Bill Apple wrote in reviewing 1956 that all of the commission's programs were "rudely interrupted in mid-year with the untimely death" of McAmis. He said McAmis was missed, particularly by him. "It was often said that if a good fight wasn't brewing, Art would find one just to keep the excitement going. Maybe that is what I loved about the guy, for I, too, love a good fight."

---

*Voting to oust Hulsey were Commissioners Rick Hampton of Stuttgart, Sullivan of Burdette, Fay Wright of North Little Rock, and Hilary Jones of Pruitt. Against were Commissioners Mike Mahony of El Dorado, Henry Moore III of Texarkana, and Dr. George Cole of Fayetteville.

Left, **T. Arthur. McAmis**

McAmis served as the director of Arkansas Game and Fish Commission from 1945 to 1956.

Arkansas Game and Fish Commission.

Right, **Nelson Cox**

Cox served as the director of the Arkansas Game and Fish Commission from 1956 to 1965.

Arkansas Game and Fish Commission.

Immediately after McAmis's funeral on May 8, 1956, the commission convened, and it appeared briefly that it had named two men with equal authority to succeed McAmis—Nelson Cox and Trusten H. "Trut" Holder. Commission chairman F. H. "Swede" McCormick of Stuttgart was forced to clarify the situation the next day. He specified that Cox had been named director, while Holder, the longtime coordinator of the commission's federal aid program, had been designated director of operations.

According to a *Gazette* article printed May 9, 1956, McCormick said the confusion resulted from "an unwarranted assumption" that Cox and Holder would have divided authority and equal status. "There can't be but one head to an organization like ours, and the commission voted unanimously for Cox to be the director and in over-all charge," McCormick explained. The *Gazette* commended the commission, saying it had avoided a "two-headed administrative monster" by clarifying the situation.

Cox served in the state house of representatives from Ouachita County in 1941 and 1943 and was chief clerk in the house during the 1953 and 1955 sessions, even after he joined the commission in 1953 as assistant director and public relations officer. Previously he was Highway Department purchasing agent during Gov. Ben T. Laney's tenure, before the Mack-Blackwell Amendment. Gus Albright, wildlife columnist for the *Gazette*, sympathized with the new director. In his column, May 13, 1956, he said,

> Misunderstandings, due partly to the public's lack of information, create many headaches for game and fish administrators. However, there are numerous other causes of controversy. Selfish interests want to be satisfied—to heck with everything else. Pressure groups search every opening to apply the wedge. Political influence contributes its share of discomfort.

Resident hunting license, 1931–32

On the back of the license were these tips on "How You Can Help": "1. Avoid forest fires. 2. Kill predacious species of hawks and owls, also stray cats and crows. 3. Observe the law and insist that others do so. 4. Encourage the growing of food and cover. 5. Do not destroy den trees. 6. Pen your dogs during the nesting season. 7. Join your local Game Protective Organization. 8. Report violations immediately. 9. Do not feel that the buying of this license completely discharges your obligation to GAME and FISH."

Arkansas Game and Fish Commission.

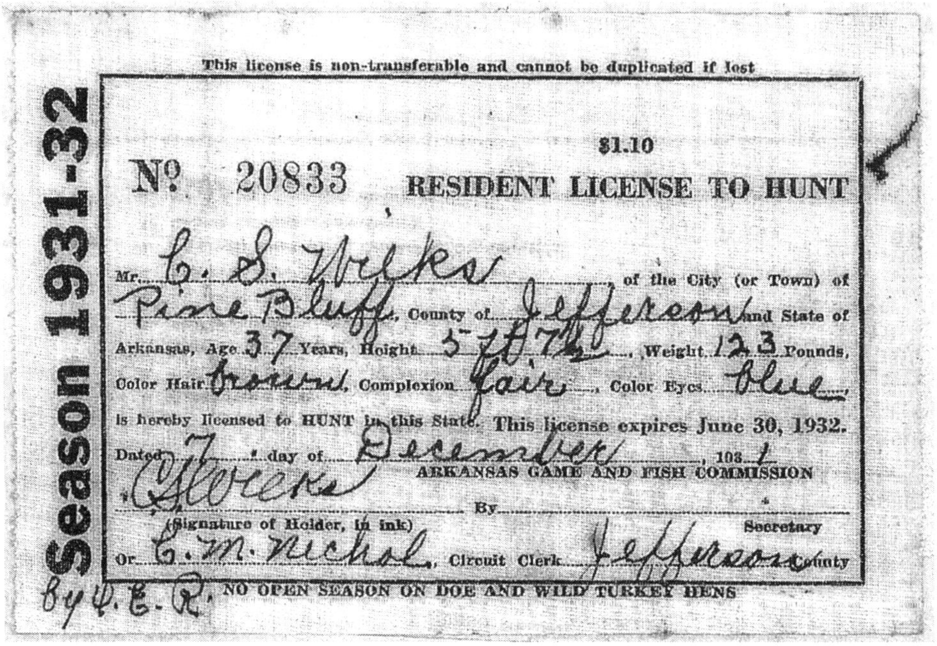

This license is non-transferable and cannot be duplicated if lost

Season 1931-32

$1.10

Nº 20833 RESIDENT LICENSE TO HUNT

Mr. C. S. Weeks of the City (or Town) of Pine Bluff, County of Jefferson and State of Arkansas, Age 37 Years, Height 5 ft 7½ Weight 123 Pounds, Color Hair brown, Complexion fair, Color Eyes blue is hereby licensed to HUNT in this State. This license expires June 30, 1932.

Dated 7 day of December, 1931

C S Weeks (Signature of Holder, in ink)

ARKANSAS GAME AND FISH COMMISSION

By............ Secretary

Or C. M. Nichol, Circuit Clerk Jefferson County

By W. E. R.

NO OPEN SEASON ON DOE AND WILD TURKEY HENS

Cox had the connections to keep the commission out of political trouble and the limelight, which is what happened. Cox resigned in February 1965, about the time Gov. Orval Faubus decided he was not going to seek a seventh two-year term as governor.

When the new Game and Fish Commission came into being in 1945, resident fishing and hunting licenses cost $1.50 each. Total income in the 1944–45 fiscal year ending June 30, 1945, was $315,123.00. Expenditures were $300,788.00. When this surplus was added to previous ones, the new commission had a total of $793,783.00 to spend. But the new agency held a tight rein on spending, keeping it to $332,818.00 and $472,490.00 in its first two fiscal years, respectively. At the same time, income shot up to $574,14.007 and $803,118.00 in the same budget years, reflecting the return of GIs who fought in World War II (Taylor 1957). No similar income increases in terms of percentages occurred again until 1954–55, when Arkansas soldiers, sailors, and Marines were returning from Korean War duty. Meanwhile, Arkansas was experiencing the most severe population loss in its history—6.5 percent during the 1950s.

The much-publicized "baby boom" was at full throttle as the decade of the 1950s opened, but demographers were looking far ahead and seeing other statistics that would influence how Americans lived in subsequent decades. Arkansas's first recognition of this demographic trend came in 1951 when the legislature passed Act 70, which provided that any resident sixty-five years or older could receive free hunting and fishing licenses. The law did not apply to commercial licenses, and it did not "repeal, alter or modify the provisions of any existing special law under the terms of which a county game commission has been created." The commission quickly established its independence on this issue when its attorney said Act 70 was invalid because Amendment 35 gave the agency "exclusive power and authority to issue licenses and permits." Amendment 35 stated further that resident hunting and fishing licenses should be $1.50 annually and "shall not exceed this amount unless a higher license fee is authorized by an act of the legislature." In other words, the General Assembly could raise the fees but not lower them or make them free.

Similarly, the commission's independence was solidified by other legal opinions. For example, the attorney general, in an opinion requested by Rep. Charlie G. Johnston of Craighead County, stated a proposed bill to outlaw "the killing and trapping of wild game on posted lands would be unconstitutional because the Arkansas Game and Fish Commission has authority over such matters through a constitutional amendment."

The first run at getting a resident license fee increase was made in 1955. A White County representative dropped House Bill 561 in the hopper with only two weeks remaining in the legislative session. It proposed to raise the resident hunting and fishing licenses to $2.50 each. A bill introduced that late in the session might be viewed only as a trial balloon. However, it gave Apple the opportunity to declare in his *Democrat* column (February 27, 1955), "There is a long gap between the days of the 10-cent pack of cigarettes and the 14-cent pound of steak, but we are still riding along on the same depression price of a dollar and a half hunting or fishing license."

The bill contained a provision exempting all persons sixty-five years of age or older from license fees. Bill Apple said this should meet with the approval of many legislators who campaigned on the promise to bring about such a revision. This section of the bill, he wrote, was for the benefit of persons who presumably have reached retirement age with a limited income and who would enjoy hunting and fishing in retirement free of charge. The most important aspect of the bill in Apple's view was the effect it would have

> on equalizing nonresident license fees and bringing them in line with those of our neighboring states. For this reason, these increases are proportionately larger and would jump the annual nonresident fishing license from $5 to $10 and the 10-day trip license from $2 to $3. Under the present law, the annual nonresident hunting license for small game is $15 and for large game is $25. Under the new bill, the annual license fee for all type game would be $25 with the special provision for seven-day hunting privileges requiring the lesser $15 license.

The commission got its one-dollar license increase in 1957 with little opposition. But part of the act went awry when Attorney General Bruce Bennett said the provision exempting senior citizens from the license fee was invalid. Bennett's opinion prompted Apple to write on September 28, 1958, "It is unfortunate that on occasion the hopes of a certain group may be built up and then the props jerked right out from under them." Had it been the General Assembly's intention to favor those age 65 or over, he said, the legislators

> would have included it in that act and permitted that age group to buy licenses at the old rate of $1.50 but no less. As the subject stands, our legislators have voted in favor of the increased license fees applying to everyone of 16 years of age or over; the attorney general's office has declared the Act of 1951 invalid; and it has been established that the conditions of Amendment 35 prevail.
>
> It seems that the . . . Commission will be within their rights in continuing their practice of requiring licenses of everyone of 16 years or over.

In 1975, the General Assembly handled the situation precisely as Apple outlined in order to give those over sixty-five years of age a break. The resident hunting and license fees for those between sixteen and sixty-five were raised to $5.00 each. Those sixty-five years or older paid $1.50 each for their licenses. Gov. David Pryor let the bill become law without his signature.

Subsequently, those who reached sixty-five years of age were made eligible to buy "lifetime" licenses, paying a one-time fee equaling a regular resident hunting, fishing or combination license. Some seventy-two thousand of these licenses were sold by 1993.

Like every other government agency, the commission was squeezed in the mid-1970s by double-digit inflation rates, driven primarily by skyrocketing energy and fuel prices. Its next resident license fee increase came in 1977 license were raised to $7.50 each.

In 1977 and 1979, the legislature began extending special privileges to other groups. The General Assembly passed laws allowing blind and 100 percent disabled persons to obtain permanent licenses for $3.50, but the commission suspended this in 1992.

Resident hunting and fishing licenses now cost $10.50 each. A $3.00 increase was obtained in 1983, in part through bargaining by Commissioner Hays Sullivan. In 1995, the General Assembly gave the commission a $1.00 license fee increase for two years because a proposed constitutional amendment was kicked off the 1994 general election ballot. (See appendixes 3 and 4. Appendix 3 depicts the history of commission license fees. Appendix 4 presents the commission's income and expenditures annually since 1945.)

Over the years, Amendment 35 stabilized the commission's personnel situation —so much so the agency occasionally announces the retirement of individuals who spent their entire adult lives working for wildlife. For example, Jasper "Doc" Freeman of Poughkeepsie retired in 1995 as enforcement supervisor for Sharp, Fulton, Independence, Lawrence, Jackson, and Cleburne Counties; he started his career as a wildlife officer in 1963. Larry Red joined the agency the same year and retired in November 1994 after spending thirty-one years at the Centerton Fish Hatchery in Benton County.

It also appears Amendment 35 has minimized legislative interference with wildlife management. On rare occasions, it is necessary for the General Assembly to exercise its authority, and every now and then, one or more of its members disagree with the agency about a specific issue and have the political muscle to translate their opinion into law.

It took legislative acts to declare all state parks bird sanctuaries in 1957 and then, ten years later, to make the entire state a bird sanctuary. The General Assembly also backed the commission in 1965 by making it unlawful to headlight deer. Also in 1965, the legislature tried to help the ailing mussel industry by exempting shells from the state severance tax.

The General Assembly either helped the commission out of controversial no-win situations or indicated its wishes or expressed its anger with the agency by passing several measures. In 1951, it told the commission forcefully in a resolution that it wanted the impoundment the agency was constructing to be named "Lake Conway." In 1959, it authorized local opinion elections on whether killing does should be allowed to continue in specified areas. In 1969, it prohibited the enforcement of any regulation regarding dogs running at large until a two-year study was made to establish the public necessity for such a rule, and "to provide safeguards for the protection of constitutional rights of individuals with respect to the use of their private property."

In 1975, the legislature required the commission to make an environmental impact statement whenever it proposes to cut timber, including selective cutting, on its lands, and to file a copy of the statement with the Natural Heritage Commission. In 1977, it prohibited hunters from establishing deer camps on highway rights of way other than at sites designed for public overnight camping by the state Highway Commission.

Politically, one of the longest-running and bitterest battles took place during the tenure of Gov. Winthrop Rockefeller. Rockefeller made no secret of his feelings about the commission when he was campaigning in 1966. There seemed to be, he said, "a growing tendency on the part of the commission to make decisions and appropriations without even bothering to explain the reasons to the people involved. During the past several years, the Arkansas Game and Fish Commission has evolved into what amounts to a private club." He accused the commission of violating the spirit of Amendment 35. After gaining office, Rockefeller asked the commissioners to resign, but they adamantly refused. Subsequently, the governor formally charged three commissioners with misconduct and had hearings held after the Arkansas Supreme Court ruled he could use this method to establish cause for removal. Commissioner Raymond Farris of Biscoe, the target of nine specific charges of wrongdoing by Rockefeller, was removed from office by Rockefeller on November 4, 1969. Farris appealed on November 28, 1969.

When Director Cox had resigned in 1965, the commission had turned to a businessman for leadership. A Mountain Home auto dealer, Hugh A. Hackler, had served in the Arkansas House of Representatives and and had been a Game and Fish commissioner from 1946 to 1953. Rockefeller wanted Hackler to resign, too, but Hackler held fast. Despite his previous service in the legislature, Hackler also had encountered problems with the Democratic-dominated General Assembly that was not inclined to support Rockefeller, a Republican. Hackler kept the commission's confidence and did not submit his resignation as director until early May 1969. Under such circumstances, it was obvious his decision to leave the commission was voluntary, but Rockefeller grumbled publicly that the action was "long overdue."

Hackler was replaced by Andrew Hulsey, who had been assistant director since March 1968. Hulsey had been with the commission more than sixteen years, nine of them as head of the Fisheries Division, when he was tapped to be Hackler's assistant. A graduate of the University of Arkansas with a master's degree in zoology (limnology), Hulsey was the commission's first director with a professional wildlife background. When he became assistant director, the *Arkansas Gazette* reported Hulsey was recognized "as one of the leading fresh water fisheries authorities in the United States and has written papers for numerous scientific publications." As are many scientists, Hulsey was apolitical. By all accounts, he ran the commission as a professional could be expected—with a strong, meticulous hand that avoided political interference.

When Hulsey was fired in August 1979, Gus Albright of the *Gazette* lamented the action, calling it a setback for Amendment 35. On September 14 that year, Albright wrote that

> Hulsey became director at a difficult time. Morale had hit bottom. There was little evidence of discipline. Commissioners were split. Many employees were unhappy. Politics had a stronger hold than ever.
>
> Hulsey literally dedicated his life to putting things back together and molding a first-class organization. He stepped on a few toes. (So will the next director if he is worth his salt).

Left, **Hugh A. Hackler**

Hackler served as the director of the Arkansas Game and Fish Commission from 1965 to 1969.

Arkansas Game and Fish Commission.

Right, **Andrew H. Hulsey**

Hulsey served as the director of the Arkansas Game and Fish Commission from 1969 to 1979.

Arkansas Game and Fish Commission.

Later, Commissioner Fay Wright said Hulsey was fired because of "long-standing dissatisfaction" on the part of some commissioners with the director's "overemphasis on certain programs" and the way he "handled things" (Hays 1969).

By the end of 1979, Hulsey was replaced by another professional biologist, Steve N. Wilson, an environmental reviewer for the Highway Department, which was at the time headed by Commissioner Wright's cousin, Henry Gray, himself a former Game and Fish Commission employee. Early in Wilson's tenure, the commission rarely made headlines. One noteworthy exception occurred in 1982 when the agency was accused of having purged nearly a dozen biologists from its staff because they had disagreed with various commission decisions. This proved to be ironic, however because the record shows Wilson, a Batesville native with a master's degree, brought a commitment to "professionalism" to the commission.

From day one at the commission, Wilson said he put the greatest emphasis on assembling a staff "of the highest quality from the education and training standpoint and productivity." For example, Doyle Shook of the Wildlife Management Division advised the commission on February 10, 1987, its timber operation was "comparable to that of one of the major timber industries in the state." To make it more efficient, he said he needed to buy six computers and software, so district foresters could analyze and report forestry inventory, prepare management prescriptions, implement treatments, and do general record-keeping. He received unanimous commission approval.

Wildlife officers (formerly called game wardens) were not neglected in the drive for modern equipment and higher standards. Cadets received extensive law enforcement training at a center the Game and Fish Commission established in 1985 at Mayflower. And in April 1986, the commission was advised the agency's mobile radios were being monitored more and more by game-law violators, lessening chances they would be caught. Permission was given immediately and unanimously for the Enforcement Division to buy 150 "state-of-the-art" radios capable of operating through a repeater or directly vehicle-to-vehicle over short distances. The cost was ninety thousand dollars.

In 1989, the General Assembly enacted a law giving wildlife officers full police powers. Later that year, Enforcement Division chief Loren Hitchcock presented new and more stringent criteria for hiring and promoting wildlife officers. The commission accepted all but one; it balked at requiring entry-level wildlife officers to have college degrees. The commission later approved this recommendation, with exceptions given to those with some college and with job-related experience. In 1992, wildlife officers became the first law enforcement officials in Arkansas to be armed with semiautomatic handguns—sixteen-shot, .40-caliber Glocks that replaced six-shot, .357 magnums.

Steve N. Wilson

Wilson has served as the director of the Arkansas Game and Fish Commission since 1979.

Arkansas Game and Fish Commission.

Fulfilling a long-time Wilson goal, the 1991 commission appropriation bill contained special language allowing incentive pay for staff members who become certified professional biologists. The emphasis on professionalism produced a staff that valued research and did not like to guess: "Research gives you specific information that helps you make better management decisions. Those decisions that come down with a strong research background behind them ought to be better for the game, better for the sportsman out in the field, allowing him to get a bigger return for his license dollar," Dr. Jim Johnson said in the commission's minutes.

As a result, when Dr. George Cole Jr. became Game and Fish Commission chairman, one of his first official statements in August 1986 was that a need existed for a graduate training and research program in Arkansas in wildlife and fishery management. He unveiled the possibility of a cooperative agreement between the University of Arkansas, the U.S. Fish and Wildlife Service, and the commission for a center to be located on the University of Arkansas's Fayetteville campus. Masters and doctoral degrees would be offered, making it a training ground for possible Game and Fish Commission staff members, and it would give those already employed an opportunity for in-service training. The commission said it wanted to participate.

Dr. Charles Amlaner, zoology department chairman at the University of Arkansas and the constitutional nonvoting eighth commission member, took the lead in establishing the research unit. In July 1987, the commission approved spending $108,787 as its first-year share of the funding. Even so, progress was slow, and it was not until January 1988 that Dr. Amlaner reported the go-ahead had been given to start hiring a staff. Dr. Jim Johnson became the unit's leader.

The commission began approving research proposals even before the unit was established, revealing a pent-up demand by its staff. At the same meeting at which Dr. Cole unveiled the cooperative agreement, the Wildlife Management Division said there was "insufficient data on bobwhite quail to assess effectiveness of present management techniques on populations" and won $19,885 for a university study.

Fisheries Division chief Allen Carter complained in October 1987 there was "a lack of data available on the microinvertebrates of the St. Francis Sunken Lands" and received commission support for a study. Amid growing pressure for more studies, Dr. Amlaner stepped in to gain commission approval for the policy of obtaining outside opinions on the scientific validity of all research proposals. Still the study proposals came: $20,000 for a channel catfish study on the Buffalo River in April 1990; $124,614 for a freshwater mussel survey to be conducted by Arkansas State University through the research unit in May 1990; $42,450 for research on stream gravel removal; and $47,000 for a study of the commission's twelve nursery ponds. All did not go smoothly, however, as Commissioner Perry Mikles of Booneville raised questions about how the money was being spent. Unhappy with the answers it received, the commission voted in June 1990 to stop paying any overhead or administrative costs to the research unit on future grants or contracts.

Above, **Wildlife officer checking license**

The job of Arkansas's wildlife officers eventually evolved into a career involving public relations and youth education as well as tireless law enforcement.

Arkansas Game and Fish Commission.

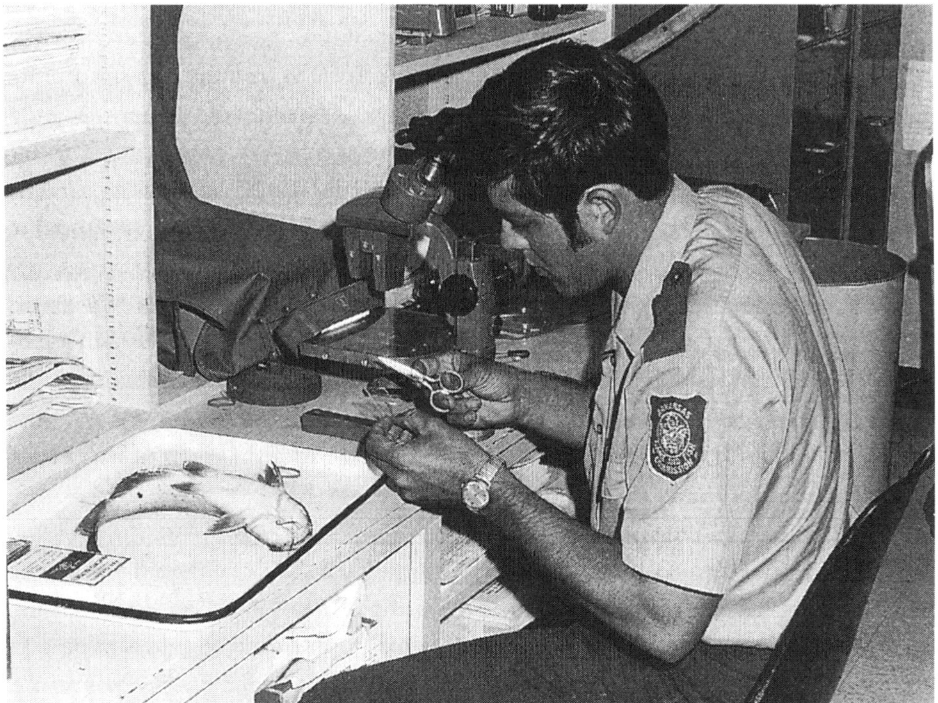

Right, **Fisheries biologist with catfish at microscope**

Wildlife and fisheries research was given increased emphasis at the Game and Fish Commission in the 1980s and 1990s.

Arkansas Game and Fish Commission.

The commission was consumed during most of the 1980s by financial concerns. Missouri had adopted a sales tax for its wildlife agency, and the commission attempted to copy the approach by getting a constitutional amendment on the general election ballot in 1984 for a one-eighth-cent sales tax like Missouri's. This was a base that would expand as the economy grew. It was projected the tax would yield $18 million in its first year.

The tax failed at the polls, however, by a vote of 350,276 or 44.66 percent for and 434,114 or 55.34 percent against. Craighead County, which had the lowest annual deer harvest in the state the previous season, was the commission's champion supporter, giving the tax 63.65 percent approval. The tax also carried in Baxter, 50.15 percent; Benton, 52.45 percent; Clay, 51.55 percent; Crittenden, 52.59 percent; Jackson, 51.50 percent; Mississippi, 51.97 percent; Poinsett, 56.65 percent; Sharp, 50.99 percent; Washington, 54.04 percent; and Woodruff, 56.39 percent. The tax came close to winning in Randolph County as well, garnering 49.45 percent. The area with the least wildlife habitat—northeast Arkansas—proved to be wildlife's strongest supporter.

Although the sales tax was defeated in 1984, Arkansas Wildlife Federation leaders and other sportsmen and conservationists were willing to try again and formed a partnership with the commission called Arkansans for Nature to gather petition signatures to get the issue on the 1986 general-election ballot. The effort was aborted when the state Department of Human Services (DHS) said the federal government would cut off food-stamp funds to Arkansas unless the sales tax on food-stamp purchases was removed. The commission's proposed sales tax would only exacerbate the situation, DHS said. The commission voluntarily withdrew the amendment from consideration.

The commission knew the increase in sales of hunting and fishing licenses was lagging behind population growth in Arkansas. In May 1986, Fiscal Division chief Ron Byrns predicted by June 1990 the commission would have a $14 million shortfall. To operate efficiently, he said, the commission's balance should not drop below $3 million at any point. The budget being prepared for the 1987 General Assembly would bring the commission down to the minimum $3 million level by June 1988. Byrns appeared again before the commission in June 1986. This time he presented a proposed budget for the General Assembly of $20,245,800. The $14 million shortfall by 1990 he had forecast was down to $8.5 million by virtue of cutting thirty staff positions and curtailing land acquisitions and various projects. This set the stage for what eventually became a showdown between the Game and Fish Commission and the General Assembly.

A commission subcommittee recommended that the 1987 General Assembly be asked for a $1 increase in resident fishing licenses (to $11.50), a $2.00 increase in resident hunting licenses (to $12.50) and a $7.25 increase in the resident sportsman's permit to $25.00. Wilson said these changes, combined with higher non-resident fees already approved by the commission, reinstatement of a fur taker's permit, and a $1,000 lifetime hunting and fishing license, would yield about $2.2 million annually compared to the $18 million a one-eighth-cent sales tax would have produced.

The subcommittee subsequently revised its recommendation to a four dollar increase in resident hunting and fishing licenses. At its February 16 meeting, the commission accepted the subcommittee recommendation, and a bill was introduced, sponsored by state senator Clarence Bell of Parkin. The commission said it might raise the sportsman's permit to twenty-five dollars as well, believing this would not require legislative approval.

Bell's bill appeared to be progressing smoothly in the legislative session until Mahlon Martin, director of the Department of Finance and Administration, told the Senate Revenue and Taxation Committee that Gov. Bill Clinton felt a 38 percent license fee increase was "more than he thinks is appropriate at this time." The committee immediately lost its quorum and could not vote on the measure.

Bell eventually withdrew the bill, explaining, "I don't believe anything will pass for Game and Fish." He said the problem was "legislative resentment about the independent status" of the commission under the constitution.

This was an unresolved situation when the commission convened March 16, 1987. When the dust settled, the commission had accepted Commissioner Frank Lyon's motion that the sportsman's permit be increased from $17.75 to $25.00 with both deer tags removed from the $10.50 license. This would require resident deer hunters to purchase the sportsman's permit and make the cost of Arkansas's most popular big-game sport $25.00. Fearing the political consequences, Perry Mikles sought to keep one deer tag on the resident hunting license, but the chairman broke a tie vote to defeat the effort.

The lawmakers complained they were flooded with calls from constituents who blamed them for raising license fees. In both the regular session and a special session that followed quickly, the lawmakers took several steps, including reattaching the deer tags to the $10.50 resident hunting license and rolling back the sportsman's permit to $17.75.

Act 939 of 1987 limited commission expenditures for the agency magazine to $450,000 in fiscal year 1988 and $150,000 in fiscal year 1989. However, the act authorized the commission to use additional funds generated by the sale of advertisements in the magazine, so long as total expenditures for the publication did not exceed $600,000 per fiscal year. In addition, the law put restrictions on commission contracts for professional and consultant services, required specified contracts be reviewed by the Legislative Council and all contracts be submitted monthly to the director of the Department of Finance and Administration.

At its March 1987 meeting in Booneville, the commission refused to follow the legislature's commands and set the stage for a showdown. In April, the commission heard a report from its general counsel, P. Douglas Mays, and decided by consensus that he should "seek a declaratory judgment in court on questions of the constitutionality of the appropriations act, and to employ a legal firm to assist." The commission also ordered printing of the sportsman's permit at twenty-five dollars.

On May 5, the commission filed suit in Pulaski County Chancery Court for a preliminary injunction to keep State Treasurer Jimmie Lou Fisher, State Auditor Julia Hughes Jones, and Department of Finance and Administration director Mahlon Martin from carrying out Act 939 of 1987. Prepared by Robert S. Shafer of the Friday, Eldredge, and Clark Law Firm in Little Rock, the lawsuit charged the General Assembly violated Amendment 35 to the constitution, which gives the commission "the sole and exclusive authority to control, manage, restore, conserve and regulate the wildlife resources of the State." The legislature, the complaint said, "is

***Arkansas Deer* and *Arkansas Conservationist* magazines**

The Game and Fish Commission has long used magazines to inform and educate the public. The commission's first issue of the *Arkansas Deer* was published in the summer of 1924. It carried the frontispiece slogan "Education Means Conservation," and the lead story devoted the publication to "more game, better laws, and finer dogs." The second issue in 1924 carried a new name: the *Arkansas Conservationist*. The last issue was printed in March 1931. The commission's current magazine, *Arkansas Wildlife*, began publication in fall 1967.

Arkansas Game and Fish Commission.

limited to making biennial appropriations for the operation of the commission and fixing the price, in excess of $1.50 each, of the annual resident hunting and fishing licenses. Any action beyond this violates both Amendment 35 and the Separation of Powers doctrine in the constitution."

Publishing a magazine is a proper commission function, the complaint continued, and discretionary matters concerning it, "including whether to advertise . . . and to what degree, and the use of advertising revenues, are matters delegated by Amendment 35 to the Commission." "The General Assembly may not use the appropriation power in this manner to direct the Commission in the specific performance of its duties," the lawsuit said.

Because it is an independent constitutional body, the commission argued, the General Assembly had no authority to make it seek Legislative Council "advice" before entering into any contract that extended more than twenty working days or exceeded five thousand dollars. It also said the General Assembly could not mandate that it adhere to general state purchasing laws.

Attorney General Steve Clark told the General Assembly in a memo on June 1 that virtually all disputed sections in Act 939 were unconstitutional or "constitutionally suspect" as the commission had claimed. Unless Act 939 and an earlier law were "substantially amended" during a special legislative session June 2–5, Clark said he would be unable to make a "good faith" defense of the laws before the court. The General Assembly made only two changes in Act 939. One deleted any mention of deer tags because Clark said this could be interpreted as a "bag limit," which was the commission's exclusive province. The other change involved the nonseverability clause in Act 939. This clause, which was opposite from the one contained in other appropriation laws, provided that if any part of the act were ruled unconstitutional, all of it would be illegal. This could have stripped the commission of the authority to spend any money for two years (the biennium) or force the calling of a special session to pass a new appropriations bill. The clause was deleted after Legislative Council director Marcus Halbrook opined that a court probably would order continued funding of the commission at existing levels in spite of what the legislature said.

Political from the outset, the situation became even more so when four legislators were allowed to intervene in the lawsuit. They were former house speaker John E. Miller of Melbourne (Izard County), Rep. Mike Wilson of Jacksonville (Pulaski County), Rep. Joseph K. "Jodie" Mahony II of El Dorado (Union County) and state senator Charlie Cole Chaffin of Benton (Saline County). Legislators set up a legal defense fund in Rep. Gus Wingfield's bank at Delight (Pike County). Meanwhile, the commission began selling the sportsman's permit with two deer tags for $25. It had no authority, however, to spend the $2.585 million the increase brought in, and the money stayed in the Game Protection Fund.

While the agency and the General Assembly were lining up their legal arguments, the commission made an effort to tighten its belt and streamline operations. Director Wilson announced on July 30, 1987, a reorganization and reduction-in-force had taken place to save the commission $3.5 million over the next four years. The deputy director position was eliminated, the assistant director retired, and his position was filled by Fisheries Division chief Scott Henderson.

The streamlining but "getting in touch" mode continued for several years with Henderson presenting a plan in October 1989 to establish ten regional commission offices. Selling a combination hunting and fishing license, one that included all necessary stamps and permits, through the mail had its genesis in a staff presentation to the commission on April 17, 1989, and went into effect in August 1991.

Chancellor John Earl signed an order December 11, 1987, allowing the commission to continue issuing permits and levying special fees, including the twenty-five-dollar sportsman's permit, while he said the legislature could continue to give "review and advice" to agencies on contracts before they were approved by the Department of Finance and Administration. Otherwise, he ruled, appropriating money from the Game Protection Fund was all the legislature could do because the commission "must decide how to spend it." With this, Judge Earl struck down the General Assembly attempt to alter the financing of the commission's magazine.

The legislators appealed Judge Earl's decision to the Arkansas Supreme Court, which ruled on October 10, 1988. It said Judge Earl was correct about the magazine but wrong about the license fees and about whether the General Assembly had authority to give "review and advice" on contracts.

In 1944–45, the only special permits or fees were for hunting dogs and some commercial licenses. "Since then," Associate Justice Darrell Hickman wrote in the court's opinion, "permits for deer, trout stamps, duck stamps, special licenses or permits for turkey and bear and use of permits for hunting with bows and muskets have been authorized"—on top of the $10.50 hunting license. Hunting and fishing could cost nearly $60.00 in fees and permits in 1986. Judge Hickman reduced the complexity to a simple question: "Does the legislature have the power to set the maximum license fees for residents to hunt and fish?" Six of the court's seven members said yes.

Amendment 35, Judge Hickman wrote, "cannot mean that both the legislature and the commission have the right to set maximum charges for resident hunting and fishing licenses. We cannot have the legislature saying a resident hunting license for deer costs $10.50 and the commission saying it costs $25.00. Amendment 35 clearly meant to give the legislature the power to determine the cost to an Arkansas resident for the privilege of hunting and fishing."

In the past, the supreme court ruled, the legislature could not set the license fee for fox-hunting dogs, could not grant free hunting and fishing licenses to those over sixty-five, and could not appropriate money from the Game Protection Fund to pay bounties on wolves. All of these, Judge Hickman observed,

> limited the legislature to the express grant of power in Amendment 35. Using that same standard for the commission, we hold it cannot intrude on the legislature's clear prerogative to set the fees in question. The commission can issue licenses and permits for residents as long as their cost does not exceed the maximum set by the legislature.

And the 1987 legislative limit, he noted, was $10.50 for a resident hunting or fishing license plus one or any combination of special fees or permits not exceeding $7.25. The $17.75 maximum did not include a $5.50 duck stamp or a $5.00 trout stamp.

Associate Justice Steele Hays disagreed with the majority. He wrote,

> The power to issue licenses and permits is located in the same paragraph which gives the Commission the authority to regulate bag limits, to regulate the manner of taking game and fish and furbearing animals, to divide the State into zones, to regulate the seasons, and to fix the penalties for violations. All the enumerated exclusive powers granted to the Commission deal with their powers of regulation, conservation, and management of our fish and game resources. Therefore, the power to issue licenses and permits was clearly viewed by this statute as another method of controlling and stocking our game population. Although the licenses and permits are admittedly a part of the revenue raising scheme, their primary purpose is inextricably related to game and fish management.

The majority's reasoning, Judge Hays continued, ignored the management function of special licenses and stamps that gives the commission "the flexibility in controlling and stocking the game resources of this State." It was legal for the legislature to set the basic $10.50 license fee, he said, but it had no right to limit the commission's power to issue special licenses and permits by restricting their total cost to $7.25.

A subcommittee of the Legislative Council reviews contracts planned by state agencies. The subcommittee at this time was made up of some of the General Assembly's most senior and influential lawmakers. A Department of Finance and Administration official testified before Judge Earl that, as a matter of policy, his agency did not allow a contract to be executed until approved by the subcommittee. In declaring this practice unconstitutional, Judge Hickman wrote for the court, "It is unreasonable to expect any state agency to defy such a body which has the power to determine the well-being" by later passing on its budget every two years. The "advice" offered by the subcommittee, the court said, "is tantamount to a legislative order on how to execute a contract." The supreme court remanded the case to Judge Earl.

Director Wilson said the ruling meant his agency would have to go to the legislature every two years for approval of its entire fee structure. Given the "tradeoffs" that inevitably occur during politically charged legislative sessions, he said, "that's a negative." Meeting on November 21, 1988, the commission dutifully reduced the fees in accord with the legislature's desire, and Fiscal Division chief Ray Sebren explained a process would be developed to refund the excess fees collected.

The trauma of the 1987–88 battle with the General Assembly forced the commission to reevaluate its image. Amid complaints that it had become "arrogant," the commission decided it needed to reach out to its constituency and instituted the practice of holding "public meetings" annually throughout the state. The commission also agreed to a policy of adopting hunting regulations for two-year periods to meet complaints the rules changed too often.

The 1989 General Assembly enacted legislation giving the commission the $25.00 sportsman's permit fee it wanted, but not until Rep. Gus Wingfield of Delight had staged a losing battle to place two deer tags rather than one on the basic $10.50 resident hunting license, which remained unchanged at that fee. Much of the legislative session was devoted to wrangling about a refund system and about how the money not claimed by hunters would be spent. Judge Earl ended the disagreement by ordering the General Assembly to appropriate any leftover funds for "hardwood reforestation, habitat rehabilitation and land acquisition or capital improvements thereon."

The amounts allotted to each were changed when the legislature convened in special session in October 1989. For example, in January 1989, the General Assembly said the commission could spend up to $1 million on land acquisition. In the interim, tracts of land some key legislators really wanted to see the commission acquire became available, and the total was raised to $2.1 million at the special session. About $1.6 million of this was to be combined with donations from Ducks Unlimited and others to buy sixty-seven hundred acres called Blackwell Bottoms near Morrilton. Some $450,000 was reserved for hardwood reforestation. About $500,000 was envisioned originally for capital improvements, but only $200,000 of it survived with new firing ranges in mind. Of the $500,000 specified for "habitat rehabilitation," $250,000 to $200,000 was to be spent on five areas intensively managed for quail.

With the air cleared by what Justice Hickman called the "legal showdown" between the legislature and the commission, many assumed the 1991 General Assembly session might be better for the wildlife agency. This was a false assumption because relations between the two bodies plummeted to a historic low. The commission entered the session with a request for a 5 percent excise tax on hunting and fishing equipment, primarily those goods on which sportsmen already paid a federal excise tax to support the Dingell-Johnson Act and the Pittman-Robertson Act. Rep. Bobby Hogue of Jonesboro was the bill's sponsor. Although the House Revenue and

Taxation Committee gave the measure a twelve to six "do pass" recommendation, Hogue soon withdrew the bill because he said he was encountering "growing opposition" to it from a legislature that already had passed a half-cent sales tax and had put the entire state four-and-one-half-cent sales tax for the first time on used cars, trailers, mobile homes, and aircraft.

Hogue raised the possibility of increasing the $10.50 fee for hunting and fishing licenses to $12.00 and for the sportsman's permit by $2.00 to $27.00. At a special meeting February 4, Commissioner Mike Cornwell of Danville vigorously opposed any license increase, but opinions from Attorney General Winston Bryant ruling out any possibility the agency could get general revenues under Amendment 35 led Hogue to amend his bill to provide for license increases anyway. Again, Hogue's bill made it out of the House Revenue and Taxation Committee only to encounter stiff opposition from what he described as a "deep reservoir of resentment" toward the commission. Needing seventy-five votes to pass the House, the bill received only forty-four with thirty-five against and five abstaining. Hogue had to abandon this approach, too.

A group of legislators—almost all supporters of the commission—met during the session in hopes of finding short-term and long-term solutions to the commission's funding problems. It came up empty-handed. This led, however, to a law creating a special legislative committee to study the situation further and make proposals to the 1993 General Assembly.

Game and Fish Commission publications from 1993 through 1995 show the agency carried out a serious spending shutdown during this period. Regulations were fine-tuned, a few parcels were acquired to be added to existing wildlife management areas, and some boat access ramps and two small regional offices were built. Otherwise, few major studies were undertaken and no sizeable land acquisitions were made unless private partners were involved heavily in helping pay for them.

For example, the Arkansas Chapter of the National Wild Turkey Federation donated substantially to four-year studies in the Ozark Mountains begun in 1991 and the Ouachita Mountains in 1993 to learn more about turkey mortality and reproduction, effects of hunting on turkey numbers, habitat for successful nesting, genetic differences, and improving management techniques. Driving this effort was the knowledge Missouri had started its restocking program about the same time as Arkansas, but hunters in the Show-Me State, with a shorter season, were bagging fifty thousand turkeys a year compared to the ninety-five hundred in Arkansas.

Perhaps the ultimate in financing complexity was reached in March 1995 when the 1,377-acre Robinwood Tract was added to Wattensaw Wildlife Management Area, making it 19,068 acres in size. Bought originally as a potential power-plant site, Arkansas Power and Light Company had wanted to sell Robinwood for years but the Game and Fish Commission did not have the money to pay the appraised price of $965,000.

This was solved when Arkansas Power and Light and its parent holding company, Entergy, donated two-thirds of Robinwood to the University of Arkansas and one-third of it to the Nature Conservancy, a private, nonprofit organization with a field office in Arkansas. The conservancy immediately donated its one-third to the Game and Fish Foundation, a private group established in 1982 to raise funds for the commission. The foundation then paid the University of Arkansas six hundred thousand dollars for its two-thirds of the property. The university divided the money among several Centers of Excellence within its statewide system. The Game and Fish Foundation's director, Steve Smith, explained his group's six hundred thousand dollars came from several sources, including proceeds from three Arkansas Outdoor Hall of Fame banquets and money from a court-ordered environmental damage settlement of a citizens lawsuit brought by the Arkansas Wildlife Federation.

The belt tightening in the early 1990s could not have come at a more unfortunate time because game and fish resource leaders across the country had awakened to the fact they were faced with unprecedented challenges wrought by demographic and cultural changes in the country.

Director Steve Wilson spoke to the changes in August 1991 in a major address before the Arkansas Wildlife Federation. One of every five Arkansans over the age of sixteen hunts, double the national average. However, the sale of Arkansas hunting licenses peaked in the 1970s and has declined and flattened since, he said. Nationally, the sale of licenses is flat or continues to decline as well. Half of all marriages in the United States fail; the fastest growing segment in society is the single-parent household, and it usually is headed by a female. Few women hunt, he said, and there are no males in many homes to teach children to hunt and fish. Further, he said, single-parent females tend to be fearful of guns or are opposed to hunting. A factor in hunting's earlier popularity was the "rural setting" that made room to roam easily accessible. Census data show Arkansas is becoming increasingly urban, and the land base for hunting is shrinking because of posting, leasing, or both. As a result, he predicted, the number of hunters will continue to decline. Because most state wildlife agencies like his rely on license fees for revenues, "we will be challenged just to maintain what we have," Wilson said.

Wildlife and the agencies responsible for its protection and management find themselves buffeted by yet another manifestation of modern society's fragmentation, single-interest groups—those who hunt turkeys and do not care about deer or vice versa, those who bowhunt and would not touch a gun, muzzleloading enthusiasts, anglers who glory in bass but sneer at flyfishers in a trout stream (or vice versa). These and others come together in specialty groups. To promote their own agendas in a world rife with competition for attention and dollars, persons with like pursuits organize so that they can exert as much influence as possible.

Sportsmen also recognize in many cases money is necessary to carry out their particular agenda, making it even more difficult for a wildlife agency to achieve balance among competing interests. Evidence of this modern phenomenon is found throughout Game and Fish Commission meeting minutes in the late 1980s and the 1990s.

The commission did what its pocketbook allowed to meet the needs of a changing society, such as holding workshops for women interested in learning about hunting and giving preference to youths who had not reached their sixteenth birthday by August 1 when it began selling special permits in 1993 for antlerless deer in areas where herd population needed reducing.

It was against this background that Arkansas attorney general Winston Bryant repeatedly told the legislative study group created in 1991 that Amendment 35 prohibited the Game and Fish Commission from receiving any general state revenues; the agency had to live on license fee income, fines paid by violators, and federal funds from taxes sportsmen paid on equipment.

The study committee decided it had no choice but to ask the 1993 General Assembly to put a constitutional amendment on the 1994 general election ballot that would give the Game and Fish Commission some financial relief. The committee's choice was to repeat what was tried in 1984—a one-eighth-cent sales tax increase. However, co-chairman Sen. Allen Gordon of Morrilton heard even some commission supporters say they were unwilling to more than double the agency's budget in a single year, particularly when the state Parks and Tourism Department had documented needs for $12 million in "new money" in each of the next ten years to make major and critical repairs at state parks and to develop parks already authorized by the legislature.

Gordon was instrumental in bringing Director Wilson and Richard Davies, director of the Parks and Tourism Department, together to discuss sharing revenues from any tax increase. Other agencies wanted a share, and the General Assembly decided to split the pot this way: the Game and Fish Commission and Parks and Tourism each would get 45 percent of the revenues, 9 percent would go to the Department of Arkansas Heritage, which includes the Natural Heritage Commission, and 1 percent to Keep Arkansas Beautiful, a group that had never been funded by the legislature. The move had the effect of broadening the constituency for the amendment, and the General Assembly ordered it placed on the November 1994 ballot.

A campaign for the amendment's passage was in high gear with polls showing likely success when the supreme court struck down another amendment the General Assembly had placed on the ballot—one to legalize charity bingo—because Secretary of State Bill McCuen had advertised it improperly. This placed the one-eighth-cent conservation tax vote in jeopardy because the secretary of state advertised it the same way. Clarence Harris of Sheridan, a retired logger, and attorney David Bowden challenged the proposed amendment, and Pulaski County chancellor Ellen Brantley threw it off the ballot less than a week before the election. This was the year the supreme court, for various reasons, removed all issues—six of them—from the ballot. Voters were angry, and about 65 percent of them marked their ballots anyway. Only votes on the sales tax were counted—unofficially—in sixty-five counties; the tallies showed the measure carrying with 53 percent support.

The 1995 General Assembly responded sympathetically, and with surprising ease: the identical constitutional amendment was ordered placed on the 1996 November general election ballot. Conditions did not appear to be the same, however because a strong antigovernment, anti-tax mood had manifested itself at the polls on other matters in the interim. Campaigners for the one-eighth-cent conservation tax could only work hard and cross their fingers for an outcome that, in effect, would save Arkansas's wildlife.

Clarence Harris tried to get the amendment removed from the 1996 ballot but lost in Pulaski County Chancery Court and did not appeal, saying that opposition to the measure was growing and he would concentrate instead on beating it at the polls.

In July 1996, Arkansas had a traumatic change in government as Gov. Jim Guy Tucker resigned and was succeeded by Lt. Gov. Mike Huckabee. Huckabee, a

Republican, was an unknown quantity to many Arkansans because his political star had risen so quickly. As it turned out, he was an avid bass fisherman with a new boat he called *State Business*, and he quickly and firmly announced his support for the conservation sales tax, explaining that Arkansas must invest in maintaining its naturalness to keep its tourism industry healthy and growing.

Gov. Mike Huckabee with Arkansas River bass.

Arkansas Game and Fish Commission.

Preelection surveys consistently showed voters favoring the amendment, sometimes by as much as 69 percent. A campaign committee, co-chaired as in 1994 by Speaker of the House Hogue and Sen. Jean Edwards of Sherrill (Jefferson County), had an easier time raising funds and was fully organized earlier than two years before. Huckabee, in mid-October, provided unprecedented support and leadership with a high-profile, four-day fishing trip on the Arkansas River punctuated by press conferences and rallies designed to promote the sales tax amendment.

Late opposition developed and surged. Two northwest Arkansas–based organizations that did not file required disclosure papers concocted a wholly fabricated link between the amendment and an alleged United Nations plot to seal off part of the Ozarks for "Man in the Biosphere" research. Columnists from opposite ends of the political spectrum in Arkansas's lone statewide newspaper railed against the amendment; neither exhibited any knowledge of why it took a proposed constitutional change to overcome Amendment 35's funding restrictions on the Game and Fish Commission or any understanding that relying on the legislature to provide money for the agency would destroy the political independence voters had bestowed on it in 1944.

On the afternoon of the day after the election, the committee was able to claim victory. Nearly eight hundred thousand Arkansans had voted on the issue, and it had won by a margin of fewer than nine thousand votes. The amendment carried only thirty-one of Arkansas's seventy-five counties. The amendment received its strongest support from the urban areas—the heavily Republican "string of pearls" of Benton, Washington, Crawford, and Sebastian Counties (plus Franklin County) in northwest Arkansas; Miller County (Texarkana) and adjacent Little River County; the Little Rock metropolitan area; Jefferson County (Pine Bluff) and the rest of heavily farmed southeast Arkansas; and Craighead County (Jonesboro) and northeast Arkansas with the exception of Lawrence County. With a notable exception, the voting pattern also suggests a correlation between support the amendment and the presence of strong higher-education institutions. The exception was Pope County, which thrashed the proposal. Pope County is home to Arkansas Tech University; ironically, Arkansas Tech trains most of the state's wildlife biologists.

No matter how narrow the win, Arkansans demonstrated once again a commitment to their environment and heritage. Game and Fish Commission director Wilson, in a personal interview, described the significance of the victory this way:

> Arkansas voters have made sure the state's bountiful natural resources will be protected and preserved for generations to come.
>
> The passage of Amendment 2 at the polls is a milestone. It's the beginning of the future, . . . This election will be remembered along with the one in November 1944, when Amendment 35 was passed to create the Game and Fish Commission in its present form.
>
> It is so significant that our people have understood the need for long-term, assured financial help for the conservation agencies . . . Certainly we thank the voters. We have to follow our appreciation immediately with a plea for patience. We will have more money to work with, but we also have to do quite a bit of catching up.
>
> Many projects of the Game and Fish Commission all around the state, big projects and little ones, have been put on hold or on the back burner these past few years because of our financial crunch. Much of this delayed work is vital. We need to rework levees at some of our lakes. We need an overhaul on a bull-dozer here and a backhoe there. Some of this type work has to come first before we can launch the innovative projects we laid out in our Plan for Conservation.

Surveys of Arkansas residents in 1994 and 1996 revealed they wanted the Game and Fish Commission to spend increased money for more wildlife law enforcement, for more educational programs, more public places for hunting and fishing, and for more work with endangered wildlife species.

"We will give the people what they asked for," Wilson vowed.

12 Lake and Stream Development

Arkansas had numerous fishin' holes, but they did not satisfy the needs of sportsmen as the pace of life quickened and as new technology prompted development of more sophisticated fishing techniques. Any up-to-date fisherman in the 1930s and 1940s wanted a rod and reel instead of a pole; plugs and flies instead of minnows and worms; as well as a good-sized boat, preferably one with an outboard motor. Few persons were content anymore to sit on a stump and dangle a line in the water waiting for a fish to bite. And few lakes in the 1930s and 1940s met these demands. Unless one preferred river fishing, Arkansas River Valley residents were particularly hard put to find a lake because the Corps of Engineers had not yet built the dams making Blue Mountain, Nimrod, and Dardanelle Lakes.

One of the first men to dream of a lake near Conway was B. F. Sterner, a Faulkner County surveyor. He was so keen on it in the late 1800s that he even chose the place on Palarm Creek where he thought the dam should be (Poe 1953). Nothing but dreaming was done until one morning some forty years later when Dr. J. H. Flanagin, a Conway dentist, asked, toward the close of a Chamber of Commerce meeting, when something would be done about a lake at Conway. He was made a committee of one to pursue it and pursue it he did.

The land at the Palarm Creek site that Sterner had chosen was too swampy to farm and had few large trees remaining. As a result, it was selling for $.50 an acre. The Arkansas Wildlife Federation collected $20,000.00 to buy land, but the publicity caused the price to rise to $100.00 an acre. A statewide campaign was begun and more than fifty thousand persons contributed. With $65,000.00 in hand, the federation began buying land. But some people would not sell (Poe 1953). The Game and Fish Commission was asked to condemn the tracts whose owners balked at selling. But the commission pointed out, correctly, it did not have the power of eminent domain at that time.

Dr. Flanagin made his first visit to the new Game and Fish Commission on January 21, 1946. He again wanted the agency to condemn the property of

In the 1930s and 1940s, when these photos were taken, Arkansans fished primarily on rivers and natural lakes. In 1945, the Game and Fish Commission initiated its lake construction program to provide additional fishing opportunities in high-population areas, in areas with few natural fishing waters, and in areas where fishing had declined despite abundant fishable waters.

Arkansas Game and Fish Commission.

Bass anglers in pirogue

Catfish anglers with cats

Man with alligator gar

Lake Conway, the first and largest lake commissioned by the Arkansas Game and Fish Commission, was impounded in the wooded Palarm Creek bottoms of Faulkner County in 1948. It is still one of the state's most popular fishing spots fifty years later.

Arkansas Game and Fish Commission (left) *and Keith Sutton* (below).

Left, **Lake Conway construction**

Below, **Lake Conway crappie angler**

recalcitrant landowners. A delegation from Conway made an appearance before the commission exactly ten months later to press its request for help.

On June 16, 1947, a Conway delegation led by state representative Russell Roberts met with the commission to discuss the proper form of conveyance and land-acquisition procedure to be used for the proposed Palarm Creek lake. Neill Bohlinger, attorney for the Highway Department, met with the two groups as an adviser. This meeting produced an agreement under which the commission would make its first land condemnations with the power given it in Amendment 35. The commission formalized the agreement in September with a unanimous vote to inaugurate the Palarm Creek project "after the land had been acquired in the name of the Commission." The agency would advertise for bids for constructing the dam at a cost not to exceed $65,000. At the same meeting, the commission awarded a $25,445 contract for construction of a fifty-two-acre hatchery at Lake Hamilton.

In November 1947, the commission elaborated on its Lake Conway agreement, requiring "all commercial operators . . . to provide for adequate public access" to common areas and to provide space for boat launching and parking entirely free of charge. No permits would be issued to commercial operators on a twenty-foot strip around the lake. Further, it said private owners bordering the lake would have access at all times, and "the 20-foot safety zone would not be commercialized or in any way used to interfere with individual rights or private development."

While the Wildlife Federation was working to get the many ends of the Lake Conway project tied up, the commission was being visited regularly by spokesmen on yet another problem—the severe bank erosion occurring on Lake Chicot near Lake Village.

The first appearance of this group was recorded March 17, 1947, when state senator Ohmer C. Burnside explained the situation. The commission told Burnside's delegation to hire a "competent engineer" to draw up plans to be presented at the next meeting. When Burnside returned in July, he reported local interests were so enthusiastic they had raised $10,000 for the restoration, and an engineer had redesigned the erosion control project, lowering the cost to an estimated $67,000 to $70,000 by eliminating some riprap and revetment work. In addition to committing itself to the Lake Conway project on September 14, 1947, the commission approved the revised Lake Chicot plans and agreed to the $66,000 cost with $10,000 pledged by Lake Village interests. Its action was conditional on the U.S. Army Corps of Engineers and the U.S. Fish and Wildlife Service giving their approval.

The commission realized it was setting a precedent with the Lake Conway and Lake Chicot projects, and it hastened to adopt a policy at its December 1947 meeting. It said sponsors of any project to build a lake or to improve any water area were required "before the Commission considers such projects, to furnish all planning, engineering, and promotion work, including preliminary survey and engineering feasibility and costs, and to furnish assurance to the Commission that the cost of the easements and rights-of-way would be borne by the sponsors." It was resolved further that all costs and expenses involved in these projects, prior to being adopted by the commission, had to be borne in full by the sponsors.

Only six months passed before the commission ended the policy. In June 1948, state representative Claud Coffelt was on hand to propose construction of a dam and lake on Wolf Creek west of Decatur. He was told all lands had to be acquired and deeded to the commission without cost, and local interests would have to provide sufficient money to build the dam after getting a fifteen-thousand-dollar contributory appropriation from the agency.

The commission came to rue its decision on Lake Chicot, for the big natural lake continually had problems of various types. Many problems proved beyond the scope and capability of the commission to handle, and the Corps of Engineers finally had to move in with a $50-million pump station—part of the overall Tensas River Basin plan—completed in 1985. For example, the commission constructed an earthen dam across the upper one-fourth of Lake Chicot in early 1948 to isolate the upper lake from agricultural runoff that entered the lake from Connerly Bayou. There was damage to the dam in December 1948. Joe Burlingame, the commission's engineer, was told to inspect the condition of the road over the dam closely and to close it if conditions became hazardous. He also was directed to consult with Chicot County authorities about either turning the road over to the county for supervision and maintenance or closing it in order to protect the dam.

In September 1949, the commission learned the engineering report it had approved for Lake Chicot was "considerably in error," and it would take twenty-five thousand cubic yards of dirt to return the existing structure to a three-to-one grade. Director McAmis was told to get Corps of Engineers to help if possible. Meanwhile, bids were advertised and taken in early 1949 on the Wolf Creek project (now known as Crystal Lake) near Decatur, but the commission rejected them as too high.

In July 1949, Orval Faubus, state representative Rolla Fitch, and engineer Burlingame came before the commission to discuss an eighty-seven-acre lake they wanted built near Hindsville at a projected cost of $22,210. Displaying the organizational talents that were later to help him become governor, Faubus had the project in line with commission policy, and the agency authorized its director to advertise for bids to be opened at the next meeting. By September, the commission heard the Hindsville project "was moving along in excellent shape . . . that work was progressing favorably with some recommended corrections on the spillway area, and also . . . that the final figures on the lake size and drainage area had not been determined," but it appeared the lake area "would be approximately one half of the original estimate of 100 acres."

The commission was advised in March 1950 that Hindsville Lake was completed. Thus, even though Lake Conway was the first to be authorized by the commission, it was not the first to be built. In fact, bids on the Palarm Creek project were not taken until four months after Hindsville was completed.

Formally opened and dedicated on July 4, 1951, Lake Conway is about eight miles long with a shoreline of fifty-two miles. It covers sixty-seven hundred acres, making it the largest lake owned by the Game and Fish Commission. Joe Hogan, fisheries supervisor at the Lonoke hatchery, stocked the lake with more than 500,000 game fish. The many trees left standing when the water rose made "ideal feeding and breeding places for the fish, which like cool, shady, still water. The roots and rotting tree trunks provide plenty of worms and bugs for their food" (Poe 1953).

South Arkansas sportsmen prevailed on the commission to take on one of its toughest projects—building a lake near Fordyce to serve three counties. Eventually, the commission had to authorize its director to condemn up to one hundred acres of land, and the agency had to withstand a legal challenge to its authority. But both the sportsmen and the commission persisted until Tri-County Lake was a reality.

Requests for lakes were pouring into the commission, forcing it to expand and be more specific about its policy. It did this on August 1, 1951, decreeing

> For the present, reasonable value of waters for new fishing developments is placed at $250 per acre of water surface for lakes and $3,000 per mile for stream channel improvement. Minimum lake size should be 50 acres. Minimum length of stream channel improvement should be one mile. New fishing developments in lowland sections . . . are not to be justified at present, since it is fairly adequate as compared to the much greater present need in upland areas. [The] pool of each lake must be surrounded by a dry land strip of sufficient width to permit proper use of the lake and control of pollution or nuisance. A horizontal width of 100 feet minimum . . . [is] suggested as proper. Greater width up to about 100 yards would be desirable. The Game and Fish Commission can always sublet its rights to such strip, and control is important.

A dam-building craze hit the United States after the war, fueled in part by a fear of massive unemployment among GIs returning from the battlefields. Some dams were authorized before the war, but others won congressional approval during the 1950s in the name of flood control and water supply.

The earliest of the major federal dams to be built in Arkansas—Norfork Dam, which created Norfork Lake—was on the North Fork River, a tributary of the White near Mountain Home in Baxter County. Congress added hydroelectric generation to the purposes of this facility, and construction proceeded despite the advent of World War II on grounds that the power was needed for the war effort. Sen. John McClellan was just as interested—if not more so—in seeing that Blakely Mountain Dam (Lake Ouachita) was constructed on the Ouachita River near Hot Springs.

In addition to Norfork, the Flood Control Act of 1938 authorized construction of five lakes—Clearwater in Missouri; Water Valley in Missouri and Arkansas; and Lone Rock, Greers Ferry, and Bell Foley in Arkansas. The Flood Control Act of 1941 modified the plan for the White River Basin to include Bull Shoals and Table Rock Lakes. The Flood Control Act of 1954 further modified the plan to include Beaver Lake for flood control and power. Beaver, Table Rock, Bull Shoals, Norfork, Greers Ferry, and Clearwater Lakes were completed. Water Valley was deauthorized.

In the Arkansas River Basin, the Flood Control Act of 1938 authorized eight flood-control projects, including Blue Mountain Lake on the Petit Jean River and Lake Nimrod on the Fourche La Fave River. By far the greatest change in the Arkansas River, however, was wrought by the construction of the McClellan-Kerr Navigation System, which was authorized in the River and Harbor Act signed by President Harry Truman in 1946. Construction began in the 1950s and intensified in the 1960s. The first cargoes reached Little Rock in 1969, Fort Smith in 1970, and Catoosa, Oklahoma, in 1971. The $1.3-billion project, which includes more than $650 million invested in Arkansas, has been described as one of the largest public works endeavors ever carried out by the Corps of Engineers. It allowed inland areas of Arkansas and Oklahoma to be linked to the Mississippi, Ohio, and other navigable systems and to the ports of the world via New Orleans.

Lake Dardanelle is a major feature of the project. With a surface area of 34,300 acres at the top of its conservation pool, it extends fifty miles upstream from Dardanelle, Arkansas, to the Ozark-Jeta Taylor Lock and Dam. Public hunting is allowed on 16,276 acres along Lake Dardanelle where the Arkansas Game and Fish Commission has responsibility for wildlife management. Its efforts are concentrated on two areas, a 220-acre area near Horsehead Park and an 825-acre area near Spadra. The commission also has constructed a 146-acre, seasonally flooded reservoir for waterfowl use in the Spadra Bottoms of Lake Dardanelle.

After the 1927 flood, several flood control and power projects were authorized for construction in the Ouachita River Basin. In 1950, additional authorizations for water resource projects were combined with previously approved ones to provide a comprehensive plan for flood control, power, and other purposes for the basin upstream from the mouth of Boeuf River. New authorizations provided for construction of DeGray Lake on Caddo River, Murfreesboro Lake on Muddy Fork, improvements on Bayou Bartholomew and its tributaries, local flood protection at

Pine Bluff, and other improvements in Louisiana. Water-resource projects from earlier authorizations that are now part of the comprehensive plan include Lake Greeson, Lake Ouachita, and drainage works on the Little Missouri River and Ozan and Terre Noire Creeks.

The Arkansas portion of the Red River Basin contains four Corps of Engineers–built lakes. Millwood Lake is on Little River seven miles east of Ashdown. It was authorized by the Flood Control Act of 1946 and later modified by the Flood Control Act of 1958. Construction began in September 1961 and was completed in August 1966. DeQueen Lake is one unit in a system of seven lakes (six in operation) authorized by the Flood Control Act of 1958 to control and develop water resources in the Little River Basin. It was constructed between 1966 and 1977 across the Rolling Fork River four miles northwest of DeQueen and serves as a water supply for that city. The most controversial of the seven proposed units was the dam across the Cossatot River six miles northwest of Gillham that formed Lake Gillham in Howard and Polk Counties. The first concrete at the Gillham spillway was poured in November 1968. Conservationists waged a fight that included a lawsuit to stop the project, which did not succeed, and it went into what the Corps calls "useful operation" in May 1975. Dierks Lake is another of the seven authorized units in the Red River Basin. Its dam is on the Saline River five miles northwest of Dierks, and the reservoir lies in Howard and Sevier Counties. Construction began in June 1968 and ended in May 1975. The Corps states that "total natural resource management—the conservation of both wildlife and natural beauty" is a major concern in the operation of Dierks Lake (*Water Resources Development in Arkansas,* 1981).

The extreme southeastern part of Arkansas is in the Tensas River Basin, scene of one of the country's most recent extensive conservation battles. Project development in this basin included the recently completed construction of a large-scale pumping system to divert pesticide-laden sediment to the Mississippi River and away from Lake Chicot, a natural oxbow.

These Corps-built dams and other extensive levee, drainage, and channelization projects extending from the 1950s through the 1970s irrevocably changed the Arkansas's face and altered its ecological balance. In the eyes of some, a number have enhanced wildlife in the state, while others have been detrimental. Still others presented challenges that Arkansas's wildlife enthusiasts overcame with innovative thinking and determination.

The Corps of Engineers and the commission were not the only government entities involved in building lakes. For example, driven by the 1954–56 droughts in Arkansas, Little Rock accelerated its effort to construct a reservoir to augment its Lake Winona water supply. The city water commission settled on a site fourteen miles west of the city on Highway 10. Bill Apple, the *Arkansas Democrat* columnist, looked forward to the project, writing on July 8, 1956,

> A bright spot now looms on the horizon for the citizens of Greater Little Rock. Never in its history has the city had any close recreational water free to public access.
>
> With the building of the huge 9,000-acre municipal water supply lake on Big Maumelle, . . . those who have visualized a fine recreational area now see this possibility crystallizing.

Apple acknowledged there was a problem connected with the impoundment:

> Can the lake be used for recreation without contamination? Other cities have found that such public use is not harmful and have opened their lakes to full public use.

After reviewing some of the scientific literature available on the matter, Apple came down on the side of public use, and said further,

> The fishing potentialities [of Maumelle] are unlimited, especially if the water commission will give thought to working in coordination with the fishery biologists of the Arkansas Game Fish Commission. Initial treatment of the basin has more to do with later fish production than any amount of stocking.

But problems began to surface even before Maumelle opened, and on May 3, 1959, Apple was forced to give his fellow sportsmen a lecture:

> Some people seem to be quite excited over the temporary closing of fishing privileges on Maumelle Lake, but let's not get the "cart ahead of the horse" as we apparently are trying to do by forcing an already settled issue.
>
> We have in the making one of Arkansas' finest recreational areas, but we must not forget that consent to its use was not a foregone conclusion from the announced date of its building.
>
> The health of the consumer comes first and has definite priority over the past-time of the fishermen or the general recreational use of the lake. Our state Health Department has final authority and jurisdiction over this and all other city water supply sources.
>
> Whether the lake remains open depends entirely on the type of use and how careful we are in observing health protecting rules for otherwise violations might result in orders that would close the waters for all types of recreation.

Certainly the Game and Fish Commission had the most unusual fishing lake in Arkansas, at least for awhile. It was the lake that lay northeast of the commission's headquarters building on the capitol grounds until it was swallowed up by construction of the Multi-Agency Complex in the late 1970s. According to commission minutes, the lake was the idea of Secretary of State C. G. "Crip" Hall. He met with the agency on June 1951 while its new building was under construction and "pointed out the advantages of a lake for educational, advertising and public relations work." The lake proved a popular fishing hole for city youngsters.

The commission's lake-building days appeared to be over in the mid-1980s as a result of the financial crisis described in chapter 11. Pressures to build lakes continued, however. At the same time, a major shift occurred in the thinking of the state's organized conservation community, which concluded Arkansas had enough "flat-water" (lake) recreation opportunities, and all worthwhile free-flowing streams should be saved in order to preserve diversity. This led to a major but losing battle with Fort Smith about damming Lee Creek for a water supply, a successful effort to keep Russellville from using the North Fork of the Illinois Bayou in the Ozark National Forest as a water supply, and a struggle to thwart the Saline County Rural Development Authority's campaign to dam the North Fork of the Saline River for a water supply. In these cases, conservationists had the commission's support, or thought they did.

Lake on capitol grounds

A fishing rodeo was held to celebrate the opening of the Game and Fish Commission lake on the capitol's grounds in 1954. Hundreds of children participated.

Arkansas Game and Fish Commission.

In the interim, however, the commission decided if landowners donated enough property to constitute 25 percent of the cost of building a new fishing lake, the gift could be used as the required match to get the remaining 75 percent cost from federal funds. Informal guidelines were adopted for this process. In addition to being equal to 25 percent of the total construction costs, the donated land had to be "suitable" (i.e., not leak or have a watershed that was too large) and the proposed lake could have "no adverse effect on existing habitat."

The donation route is how Lake Jack Nolan, originally Lake Greenwood and also called Lake Wright, came to be in Sebastian County. U.S. Forest Service insistence that roads had to be relocated effectively blocked construction of a new fishing lake in Scott County because of the expense. The commission, however, forged ahead with preliminary work for a lake on Sullivan Creek in north Independence County on land to be donated by the family of state senator Steve Bell of Batesville. The project was eventually dropped when additional tests for water quality were ordered. The commission did not have funding available for such testing.

A resolution urging the commission to be cautious about accepting donated land for new fishing lakes survived a motion to table and was adopted by a two-to-one margin at the Arkansas Wildlife Federation's annual meeting in August 1990 ("G&FC Urged to Be Cautious about New Lake Proposals," 1990). Delegates from northeast Arkansas, where fishing lakes were under intense fishing pressure, opposed the resolution.

The Arkansas Wildlife Federation resolution complained that donations allowed landowners to obtain "sizable tax deductions" while increasing the value of lakeside acreage that could be sold and developed for homes. "There could be increased destruction of smallmouth bass streams and other high quality waters, and degradation of sensitive terrestrial habitat in ecologically important areas if this problem is not closely monitored and controlled," the resolution warned. It also urged the commission to establish a buffer of a half-mile or more around any lake "to eliminate the danger of future lakefront developments" and resulting pollution from septic tanks.

As to Arkansas streams, particularly those featuring smallmouth bass, an article in the commission's magazine raised alarm about them in spring 1991: "Our smallmouth streams are under siege from a variety of factors. Water quality, population and habitat degradation comes from damming, angler pressure, gravel operations, municipal waste discharges, poultry waste and stream channelization," the article said. Smallmouth bass, commission magazine editor Gregg Patterson noted, do not flourish anywhere other than in the "clean, cool, clear mountain streams and a few mountain lakes" in the Ozarks and Ouachitas. If the fish flourishes, he continued, "then Arkansas will live up to the Natural State nickname. But if the smallmouth bass is reduced to scattered remnants found only in remote areas, then we will have lost the richness that makes the state so attractive." Organized conservationists began to see the commission as having inconsistent policies as the agency juggled demands between people and the resource.

As the environmental movement gained momentum nationally and in the state in the 1960s and 1970s, some of the emphasis previously devoted to creating habitat turned to preserving what still existed. After federal dam after dam went up on Arkansas's rivers, conservationists finally said enough was enough when plans were going forward for the Water Valley Dam on the Buffalo River. The environmental movement came of age with the battle to save the Buffalo, which was led by the newly organized Ozark Society under the leadership of Dr. Neil Compton. Sportsmen were the conservationists' allies, as Apple showed on July 14, 1957, when he wrote,

> Congressman Jim Trimble is continuing his efforts toward damming the beautiful Buffalo River, in spite of the fact that the U.S. Engineers have declared the damming would be economically unjustified. Should he be successful, we will see the end of our mountain float fishing and the drowning of Arkansas' most beautiful stream that has won national acclaim, attracting visitors in countless numbers to that area.

Trimble did not win. On March 1, 1972, President Richard Nixon signed into law legislation "for the purposes of conserving an area containing unique scenic and scientific features, and preserving as a free-flowing stream an important segment of the Buffalo River in Arkansas for the benefit and enjoyment of present and future generations." On that date, the Buffalo became the country's first national river.

Bill Apple was among the first conservationists to be credited with protesting the Corps of Engineers' plan to channelize the Cache River and its principal tributary, Bayou DeView.

Writing in *Sports Afield* (April 1955), Cleveland vanDresser said,

> Picture this situation: a lush valley extending southward from the Missouri line. Through 210 miles of it, the gentle Cache River flows to where it meets the White River in southeastern Arkansas. The Cache's 210-mile length is dotted with lakes, bayous, marshes and forests, all crammed with wildlife.

> Since the dawn of American history, this valley has been part of the greatest hunting and fishing area in the United States . . . and it is slated for a drastic drainage program that will leave it almost devoid of wildlife.
>
> Fifty years ago the bottom land of Arkansas—an area of 18,000 square miles—or roughly a third of the state—was a wildlife paradise. The amount of game, mainly waterfowl, produced and supported by that natural area was fantastic.
>
> There is a physical factor that is not mentioned by the Engineers. You can't shunt water around in the interior of the country without drastic results. If an aborted Cache River dumps the load of 2,000 square miles of drained marsh land into the White River, the White will get an added load. So a definite flood hazard will be created on the lower White River.
>
> When William Apple, secretary of the Arkansas Wildlife Federation and a leading conservationist, made this situation clear to the residents of the lower White River, a scream of protest went up from that section of Arkansas.

In the early 1970s, Pratt Remmel Jr., of the now-defunct Arkansas Ecology Center; his research assistant, Gale S. Eddins; and Texarkana lawyer (later United States Judge for the Eighth Circuit) Richard S. Arnold persuaded the Environmental Defense Fund to file suit challenging the adequacy of the Corps of Engineer's twelve-page environmental impact statement on the Cache project. The Game and Fish Commission and the Arkansas Wildlife Federation entered the litigation on the side of the conservationists, representing the first time a state agency joined with private groups in a modern environmental controversy and setting an important precedent. Federal District Judge J. Smith Henley ruled for the Corps. Gov. Dale Bumpers asked the Corps to delay digging until the Eighth Circuit Court of Appeals could rule. But the Corps started work anyway on the lower end of the project near Clarendon.

"I couldn't stand by and watch a bureaucratic federal agency thumb its nose at Arkansas," Stuttgart dentist and avid sportsman Dr. Rex Hancock said, explaining why he single-handedly organized the Citizens Committee to Save the Cache River in October 1972. The committee eventually included thirty-five national organizations

Aerial view of Buffalo River camp

This photograph (circa 1959) indicates the Buffalo River was a popular destination for outdoorsmen long before its designation in 1972 as our country's first national river.

Arkansas Game and Fish Commission.

East Arkansas's Cache River basin encompasses some of the most important bottomland hardwood habitat in Arkansas. Efforts of the Corps of Engineers to channelize the Cache drew national outcry in the 1970s. Work began at the river's mouth near Clarendon, with two huge draglines running twenty-four hours a day. The project was halted, however, when Congress cut off funding in 1978. The U.S. Fish and Wildlife Service described the project as "the single most damaging project to waterfowl in the nation."

Arkansas Game and Fish Commission.

Cache River in its natural state

Cache River channelization work

and eight states in the Mississippi Flyway. Hancock challenged the Corps at every turn, spending hundreds of hours and thousands of his own dollars. Meanwhile, the Eighth Circuit Court of Appeals sent the Corps back to the drawing board to prepare a better environmental impact statement.

Congress cut off funding for the project in 1978 after the U.S. Fish and Wildlife Service called it "the single most damaging project to waterfowl in the nation." Then the Corps encountered Adlene Harrison, regional administrator of the federal Environmental Protection Agency, who refused in 1979 to approve a permit required under Section 404 of the Clean Water Act for the project.

The Fish and Wildlife Service designated part of the lower Cache River basin as a thirty-five-thousand-acre national wildlife refuge in 1980. The Nature Conservancy took the lead in acquiring land for the refuge to be turned over to government agencies, and the wildlife refuge was on the way to reality. In 1986, the Cache River/Bayou DeView area became part of a larger, more ambitious international plan to save waterfowl populations that had been dwindling rapidly as a result of drought and habitat destruction.

Game and Fish Commission director Wilson and waterfowl biologist Dr. Scott Yaich announced on April 13 and 14, 1987, the United States and Canada had signed the North American Waterfowl Management Plan (NAWMP). The linchpin of the plan was broad-based aggressive habitat protection programs such as the Quill Lakes Development Proposal on eighty-four thousand acres of breeding and staging lowland marshes in southeast Saskatchewan, Canada. The commission promptly made a requested one-hundred-thousand-dollar contribution to the Quill Lakes effort.

The plan's goal was to help ensure habitat for 62 million breeding ducks on the continent and a fall migration of more than 100 million ducks, levels comparable to the mid-1970s.

Seven months later, Dr. Yaich reviewed the NAWMP more extensively and said a joint venture had been put together for the Lower Mississippi Valley (Patterson, "Dark Days for Ducks . . . Is There Hope" 1988). It identified 386,000 acres of wetlands targeted for protection and purchase by the year 2000, including 145,000 acres in Arkansas, primarily in the Cache/Lower White River region. This acreage was to be in addition to the already protected White River National Wildlife Refuge, the newly designated Cache River National Wildlife Refuge and several Game and Fish Commission wildlife management areas.

Wilson called the plan, developed by the Fish and Wildlife Service and state natural resource agencies across the country working with private groups such as the Nature Conservancy, Ducks Unlimited, and the National Wildlife Federation, "the best thing that's happened for waterfowl since the federal Migratory Waterfowl Stamp Program began in the Dust Bowl days of the 1930s" (Wilson 1988). Yaich said the plan represented one of the last opportunities to protect an entire river corridor and its importance was unprecedented in the Mississippi Flyway. A third of the flyway's mallard population of 3.1 million wintered in Arkansas's Delta.

Under the Lower Mississippi River Valley Joint Venture, five programs were developed to obtain the additional 145,000 acres of winter duck habitat needed by improving existing management practices, habitat acquisition, and offering waterfowl management expertise to cooperating private landowners. To carry out its part of the Lower Mississippi Valley Joint Venture, Arkansas developed an implementation plan that emphasized improving private ownership management practices. Yaich explained why:

> On public lands today, we can at best provide for only 20 percent of the needs of ducks if we have a reasonably wet fall and winter. That means around 80 percent of waterfowl food and habitat needs are being met on private lands today. During dry years, these needs simply aren't met, and ducks suffer. Our goal for the year 2000 is to provide 35 percent of waterfowl needs on public lands and the remaining 65 percent on private lands. That assumes some pretty lofty goals with regards to acquisition of additional public lands and improvement of habitat on those lands. With existing revenue sources, the AGFC can't meet those needs. This reemphasizes the need for private land management. (Sutton,"What You Can Do for Ducks" 1991)

Needs under the Arkansas Implementation Plan (AIP) translated into almost one million acres of managed land and water to provide habitat for the goal of 2.9 million ducks wintering in the state by the year 2000. Private lands compose about 90 percent of the state, and the AIP goal was to have three hundred thousand acres of these managed for waterfowl.

"Our most recent estimates indicate that only about 225,000 acres of wetlands and harvested fields are managed on private land in Arkansas in a normal rainfall year," Game and Fish Commission wetlands coordinator Jon Schneider reported in May 1996. "To address this shortfall, we've developed one of the most successful efforts of the plan, the Arkansas Partners Project, a cooperative venture of the commission, Ducks Unlimited, U.S. Fish and Wildlife Service, the U.S. Natural Resources Conservation Service, and private landowners." The program provided free water-control structures and paid for tree planting in exchange for a fifteen-year management agreement with landowners. By May 1996, Schneider said the Partners Project had worked with 150 landowners, provided nearly five hundred structures and had signed management agreements on more than thirty thousand acres.

Relative to the overall North American Waterfowl Management Plan, private, state and federal partners by 1996 had spent nearly $1 billion to protect, restore and enhance more than 4.5 million acres of waterfowl habitat on the continent. In Arkansas, the Partners Project had spent more than $64 million in cash and some $21.4 million in service to acquire 118,566 acres, restore 27,558 acres and enhance 79,791 acres.

Acquiring land or working with private landowners to improve habitat for waterfowl was not the only major step taken in Arkansas to protect and preserve the resource. The Fish and Wildlife Service had determined lead poisoning, caused by waterfowl ingesting lead shot while feeding, caused the direct mortality of a conservatively estimated 1.5 million to 3.8 million waterfowl a year. Other evidence strongly indicated millions more were dying of causes indirectly related to lead poisoning (such as birds weakened by lead poisoning being more susceptible to disease or predation) or were living with permanent brain damage, impairment of normal reproduction, and other physiological disorders.

In 1976, the Fish and Wildlife Service tried to implement a phase-in program to eliminate lead shot, with steel shot as the replacement. This effort was aborted in

Woodruff County mallards

Mallards rise from a flooded crop field in Woodruff County. The Arkansas Partners Project, implemented in 1996, emphasized the importance of private-lands waterfowl management efforts in the state.

Keith Sutton.

1979 by congressional action. After years of bitter public controversy and a long public review process, the agency developed and published final guidelines for the change on July 30, 1985. The guidelines were written as minimum standards and affected all states. Those areas with the greatest known or potential problems were to switch first. The National Wildlife Federation favored immediate statewide conversion to steel shot, but it apparently stood alone in this, and the commission adopted a phase-in plan for Arkansas accelerated slightly by the federal agency. Fourteen counties converted to steel shot in the 1987–88 season, and six were added in 1988, eight in 1989, and the remaining ones in 1990 to meet the Fish and Wildlife Service's requirement that all waterfowl hunting be with steel shot by the 1991–92 season.

Congress enacted the Clean Water Act in 1972, and the new Environmental Protection Agency focused on getting states to curb pollution by regulating discharges from industry and from municipal sewage treatment facilities. Enough progress on controlling these so-called "point discharges" had been registered by the late 1980s that the nation shifted its attention to "nonpoint pollution," which comes from dispersed activities such as farming, logging, and urban runoff.

In Arkansas, water-quality concern shifted in large part from point to nonpoint sources and from lakes to streams in 1991 when John Giese, chief of the Pollution Control and Ecology Department's Environmental Preservation Division, reported some of Arkansas's premiere streams were contaminated by nitrates and fecal coliform bacteria. Runoff from chicken and hog manure spread as fertilizer was a suspected culprit. The Pollution Control and Ecology Department subsequently adopted regulations to foster better "liquid animal waste management" practices, principally on hog farms. Chicken litter disposal continued to be unregulated.

In June 1991, the Game and Fish Commission adopted a broad order directing its staff to work with other organizations in developing a program to combat growing pollution problems because its constitutional duty of managing the state's fisheries resources was "inexorably related to the continued vitality and maintenance of water quality." Giese identified the Buffalo River, Richland Creek, North Sylamore Creek, Strawberry River, Spring River, Eleven Point River, Lee Creek, Mulberry River, Kings River, Current River, Piney Creek, Hurricane Creek, Illinois Bayou, Little Red River, Salado Creek, Cadron Creek, Mountain Fork, Cossatot River, Saline River, and Second Creek as some of the Arkansas streams with contamination problems. Other streams being threatened by construction and other activities, Giese said, included the White, Little Missouri, Ouachita, L'Anguille, and Cache Rivers.

Oil was discovered in the El Dorado, Smackover, Magnolia, and Lewisville areas of south Arkansas in the early 1920s. When oil is extracted, it is accompanied by saltwater that must be separated from the petroleum. During early oil production days, most operators let the saltwater waste go where it would—across the land, destroying vegetation, or into creeks and rivers, loading them with chlorides.

World War II brought an industrial base to Arkansas in the form of several government-owned and privately operated aluminum plants. Other industries located in Arkansas to extract minerals as part of the war effort. In addition to jobs, the plants brought a legacy of pollution to the state that began reaching harmful levels in the late 1940s.

Water pollution problems were assigned at first by the General Assembly to the Game and Fish Commission. The commission was trying to counter the pollution as early as 1947 when it told Director McAmis to contact officials at the Barium and Magnet Cove plants as well as the National Lead Company near Malvern to warn them the agency was prepared to take action against them if they did not do something to stop their pollution. A year later, Director McAmis was instructed to write to mine operators along James Fork Creek in Sebastian County and give them the same warning. But McAmis knew when a problem was beyond him, the commission, and any other government agency designated to deal with it. Armed with an attorney general's opinion that outlined the limited jurisdiction that existed, McAmis advised the commission of his view, and in October 1948, the commission agreed "there was a need for an over-all pollution control measure that would bring other agencies into the picture." McAmis was instructed to help draft the legislation to bring this about. The resulting bill created the Water Pollution Control Commission in the state Health Department in 1949, forerunner of today's Department of Pollution Control and Ecology.

The Arkansas Supreme Court in 1966 held that the Hercules Powder Company of Jacksonville pay damages to commercial fisherman Kelsey Carson because it had polluted Bayou Meto. In 1979, the Environmental Protection Agency revealed that chemical plants occupying the Jacksonville site over the years contaminated the land and nearby streams, including Bayou Meto, with TCDD, the most dangerous of dioxins. The announcement touched off a battle between some Jacksonville residents and federal and state agencies about the extent and effects of the polluants and how they should be destroyed. The struggle did not end until the last of the wastes, stored in barrels, were hauled to another state in 1996 to be destroyed by incineration. Meanwhile, Bayou Meto remains polluted by steadily decreasing TCDD amounts in fish and sediment.

Other suspected carcinogenic chemicals, PCBs, were found to have contaminated the sediment and fish in Lake Pine Bluff in June 1980. Until their use was banned, PCBs were found in products ranging from carbonless paper to electrical transformers. In 1988, the Game and Fish Commission killed the fish and drained Lake Pine Bluff, after which the lake bed lay exposed for two years while nature took its course. As the bed dried and hardened, the PCBs seeped deep into the mud and the top layer crusted. After two years, the lake was refilled and restocked with fish. Total cost to the Game and Fish Commission for the cleanup and restocking was about $250,000. In 1992, Fisheries Division assistant chief Mike Armstrong reported that the cleanup had worked because PCB levels were staying well below what the Food and Drug Administration said warranted action.

In September 1992, Arkansas was rocked when it became one of twenty-eight states that had to issue advisories against consuming some fish from certain waters because of high (more than one part per million) methyl mercury levels in their flesh. Mercury can damage the human nervous system. Most advisories were for fish from waters in south Arkansas, but they reached as far north as Lake Nimrod in Yell County and Johnson Hole on the South Fork of the Little Red River. The Arkansas public immediately suspected industry was releasing the mercury, but extensive testing downstream from various plants and painstaking searches by pollution control officials found no hidden discharge pipes from factories or any other source. A governor's task force learned that experts believed the mercury may be a result of natural elements leaching from the soil or rocks or may be coming from the air (coal and waste incineration) or perhaps may be a result of a combination of the two.

Sulphur-reducing bacteria convert elemental mercury deposited in sediment into the methyl mercury that can be taken up by algae. Small fish eat the algae. They, in turn, are eaten by larger fish. The mercury builds up as fish grow. The Game and Fish Commission published pamphlets and other materials for the public about the mercury contamination, advising among other things that largemouth bass exceeding sixteen inches in length and catfish usually are the principal species affected and that young children and expectant mothers are the most susceptible to problems stemming from mercury. Testing of fish in Arkansas waters continues.

No one was more upset about what in-stream gravel mining was doing to the integrity of Arkansas's waterways than the Pollution Control and Ecology Department's John Giese. Fine particles of sand and dirt stirred up by gravel operations settle out downstream. This silt gradually coats pebbles and fills spaces between them, smothering the tiny creatures (invertebrates) living there. When the invertebrates die, so do fish because their food has been destroyed. Revealing the depth of his feeling, Giese said,

> I'd rather you go out there and dump a pesticide in the stream that kills the fish. The fish will come back after detoxification, but with sand and gravel, the stream gets imbedded, and you just have big mud flats that are sterile.

A study done by Dr. Arthur V. Brown and Madeleine M. Lyttle for the Arkansas Cooperative Fish and Wildlife Research Unit at the University of Arkansas at

Fayetteville gave Giese and others the ammunition needed to act. With a special focus on game fish, Dr. Brown and Lyttle studied the impact of gravel removal on stream ecosystems at sites in Kings River, Crooked Creek, and Illinois River. From their data, they concluded gravel mining "significantly degrades the quality of Ozark stream ecosystems." Shallow downstream flats caused by siltation "favored larger numbers of a few small-fish species" and the removal of vegetation caused fish and invertebrates to become stunted at both the mining sites and downstream. The 1993 General Assembly enacted a law banning commercial gravel mining in twenty-four streams and some of their tributaries that were designated "extraordinary resource" water bodies in Arkansas's water-quality standards and authorized Pollution Control and Ecology to draft regulations for permits to take the commodities from other waterways.

The gravel miners raised such a political furor with threats they would get the ban repealed that the Pollution Control and Eology Department's director Randall Mathis elected not to enforce it until the 1995 General Assembly could "revisit" the issue. In the interim, 1,500 persons signed petitions asking the department to designate Crooked Creek, a prime smallmouth bass stream, as an "extraordinary resource" in order to block further mining. Also, the department tried unsuccessfully to get "favorable review" from a legislative committee on regulations to control sand and gravel mining in non-extraordinary resource streams.

The three issues—the ban, the proposed regulations, and the requested designation of Crooked Creek—became hopelessly entangled and set the stage for a bitter fight that raged throughout the 1995 General Assembly session. When the smoke cleared, a bill and strategy devised by Senators Wayne Dowd of Texarkana, Vic Snyder of Little Rock, and Nick Wilson of Pocahontas, preserved the ban on commercial gravel mining from "extraordinary resource" streams and Pollution Control and Ecology's authority to adopt regulations to control the activity in other waterways. In early 1997, the department abandoned the process to name Crooked Creek an "extraordinary resource" stream.

13 Land Acquisition and Development

The old Game and Fish Commission made a stab at remedying the lack of public hunting areas by establishing one on the east side of the White River in Monroe County in 1943. However, it was not until Amendment 35 was adopted that the commission felt it had the mandate to do some of what the it had been recommending for years. The General Assembly waited two years, until 1947, to appropriate some of the funds the agency needed to carry out that mandate. Meanwhile, a group as far behind as the commission was on implementing needs identified such a long time before could not—and would not—be choosy.

The army had established two large training camps in Arkansas during World War II. One was Camp Robinson near North Little Rock and the other was Fort Chaffee at Barling near Fort Smith. When the war ended, the army found it had some surplus land on its hands. The new commission wasted no time speaking up for some of it. At its May 1946 meeting, the commission authorized member Harry Barr of Fort Smith "to proceed with the request and necessary expense in following up the acquisition of approximately 17,000 acres of the 90,000-acre tract of the Camp Chaffee area that is to be declared surplus by the government. This action was taken so as to bring this area into our refuge system and was passed unanimously." Despite the speed with which the commission moved, it was not until March 1950 it was able to approve the acquisition, along with a plan to develop the seven-thousand-acre Big Creek Narrows Lake area at a cost of $81,000. In April 1948, Director McAmis briefed the commission on action taken to acquire about 13,400 acres of Camp Robinson and for game management rights on an additional 42,000 acres. The commissioners asked McAmis to proceed "on this procurement" and to keep them advised. At its meeting two months later, the commission told

McAmis to proceed with acquiring 35,000 acres of Camp Robinson in the refuge program by agreement with the War Department and the State Adjutant General's office.

During the Depression, the federal government acquired substantial tracts of land for agricultural resettlement communities and other purposes. When the Pine Tree, Wattensaw, Marianna, and Camden Land Utilization Project Areas became available in 1951, the commission spoke out for about sixty-six thousand acres of them.

The acid test for the new commission was its first acquisition of land for a combined hunting area and refuge that was not already owned by a public entity. That test began to take shape February 16, 1948, when a committee from Ducks Unlimited and the Arkansas Wildlife Federation met with the commission to

> present a public duck shooting proposal . . . [They] pointed out the need and advisability of establishing wildlife management areas, with particular emphasis on two large tracts of land south of Stuttgart; one on Bayou Meto, embracing approximately 7,000 acres; and the other on Grand Cypress, lying partly in Jefferson County.

The committee asked the commission to take immediate steps to survey and make engineering plans for the eventual acquisition of these lands. With little or no hesitation, the commission authorized McAmis to consult with federal wildlife officials about hiring an abstractor and hydraulic engineer immediately "for the purpose of putting this plan into motion."

In May, McAmis reported he had received approval from the Fish and Wildlife Service to acquire the Bayou Meto land at not more than $10 an acre or $125,000. Moving rapidly, the commission at its meeting in June directed Trusten H. Holder, its coordinator of federal programs, to get an option on 36,500 additional acres to complete the Bayou Meto project. Plans called for flooding 10,000 acres at a pool elevation of 175 feet mean sea level.

By October, the commission was discussing what could be done in the way of some temporary development work on what it was now calling the wildlife management area in Arkansas County. For three thousand dollars, the commission was advised, a temporary structure could be installed that would impound surface water "as a possibility of assuring some duck shooting in the area this year." But the structure also would inundate some land the commission did not own, and action was deferred. In February 1949, Holder returned to the commission with a proposed amendment to purchase the Wrape Plantation as part of the Bayou Meto Wildlife Management Area. He described it as suitable for feed production and waterfowl resting, both of which were required by the federal agency. The commission gave its go-ahead. Negotiating with the various interests in the area was not easy, but the going got even tougher in September 1949 when bids for installing structures at Bayou Meto came in and were deemed "excessive" by the commission and rejected.

State senator Lawrence Blackwell appeared at the commission's August 1, 1950, meeting in hopes of hammering out a compromise between the agency and some Grand Prairie landowners who did not want to give up their land by condemnation or any other means. The commission rejected Blackwell's compromise, and the landowners went to court to challenge the commission's right to take 1,320 acres in Jefferson County to be incorporated as part of the wildlife management area.

Bayou Meto duck hunters

In 1948, the Game and Fish Commission made the first land purchases for what would eventually become Bayou Meto Wildlife Management Area. Huge flocks of wintering mallards have made this one of the most popular duck hunting areas in the state since its earliest history. The successful hunters pictured here posed in the early 1950s.

Arkansas Game and Fish Commission.

Section 8 of Amendment 35 states the commission "shall have the power to acquire by purchase, gifts, eminent domain, or otherwise, all property necessary, useful or convenient for the use of the commission in the exercise of any of its duties, and in the event the right of eminent domain is exercised, it shall be exercised in the same manner as now or hereafter provided for the exercise of eminent domain by the Highway Commission." In addition, the commission had won an earlier challenge to its eminent domain power, brought by the Wrape Stave Company, involving six thousand acres condemned for Lake Conway. Thus it appeared the commission was in good shape, and it won a lower-court decision on Bayou Meto.

However, on April 16, 1951, the commission was stunned when the Arkansas Supreme Court released a four to three decision (*Hampton v. Arkansas Game Fish Commission*) in which it reversed the lower court and said the agency did not have eminent domain powers to acquire public shooting grounds. The court cited McAmis's testimony that public shooting would be allowed on about 75 percent of the forty-thousand-acre Bayou Meto tract. It also quoted from a twenty-three-page report Holder submitted to the commission in which he asserted,

> Prior to and during the 1920s and 1930s, duck hunting was commercialized to a considerable extent by professional guides . . . The decline in the national duck population made hunters realize that ducks could no longer be killed at just any place . . . There are less than 3,000 duck hunters who really have a place to hunt, and about 50,000 would-be duck hunters in Arkansas . . . It is uncertain just how much longer most duck hunters will buy hunting licenses and duck stamps, with only a faint hope of getting to go on a good duck hunt . . . The only possibility of doing something constructive, however, is through the purchase of public hunting areas. We are about 15 years late in getting started, but not too late to take advantage of several outstanding areas that are still available.

The court concluded that

> in the final analysis, the evidence shows that this is primarily a duck hunting project and not a project to preserve wildlife . . . In the light of the evidence and the rules concerning eminent domain, as previously stated, we have no hesitance in holding that Amendment 35 does not give the commission power of eminent domain to establish a public hunting ground . . . The plan here is to kill ducks, and killing is certainly the antithesis of restoration and conservation. Neither can killing be said to be control, management, or regulation.

The difference between the Bayou Meto and the Conway cases, the court explained, was that the first and primary purpose of the lake was for the propagation of fish.

Associate Justice George Rose Smith wrote a dissent in which two others, Justice J. Seaborn Hold and Justice Minor W. Millwee, concurred. "I cannot escape the feeling that by today's decision the majority has discriminated against hunters as compared to fishermen." After all, he pointed out, the purpose of propagating fish in Lake Conway was so they could be caught by anglers. He asked how that was different from helping to propagate ducks in Bayou Meto to be shot by hunters. Judge Smith also said the majority had "confused the meaning of 'conserve' with that of 'preserve.' The latter means to protect entirely from danger, but conservation 'stresses the idea of maintenance of an existing condition.'"

Thus the commission's first effort to provide a combined public shooting ground and game refuge on land not already owned by the public was traumatic. But Holder was not deterred. Even while involved in the Bayou Meto fray, he came to the commission with plans approved by the Fish and Wildlife Service in November 1950 to use Pittman-Robertson funds to help acquire 10,000 acres at $10.00 an acre in the Black River area of Clay County. Holder was back before the commission in December 1952, seeking approval to add 1,200 acres to the Black River area at $7.50 an acre. This was followed two months later by the commission's decision to buy 1,340 acres near Petit Jean for $10.00 an acre. Eight months later, the commission approved buying another 322 acres at Petit Jean for $4,800.00. A refuge system was taking shape.

For reasons described in chapter 11, the days of dramatic land acquisitions came to a halt in 1989 for the Game and Fish Commission with the purchase of the Ed Gordon/Point Remove Wildlife Management Area. As the list of wildlife management areas suggests, however, the commission was able to enlarge its holdings in subsequent years through an unexpected partnership with the federal Farmers Home Administration (FmHA). An depression in the agricultural industry in the 1980s forced some farmers to default on their FmHA loans, and the federal agency literally ended up owning the farms. Federal law provided that, when suitable, the land could be transferred to the state for wildlife purposes. Through mid-1996, the Game and Fish Commission had received seventeen tracts totaling more than 7,500 acres (ranging in size from 78 acres to 1,503 acres) from the FmHA. Each became a wildlife management area.

Today the commission has nearly 320,000 acres of wildlife management and demonstration areas, making it first in ownership of land dedicated to wildlife among states in the southeastern United States and fourth nationally. It also leases and has cooperative agreements for another 2.1 million acres.

In early 1987, a perceptible shock struck those at a commission meeting when member Casey Jones of Pine Bluff announced he had been informed that Georgia-Pacific Corporation and others among the state's major timber companies would accelerate a program of posting their lands and leasing them to private hunting clubs. Jones assured the commission that the companies intended to make large areas available to the agency for leasing in order to keep hunting affordable to the general public. This was a sea change in the state because Arkansans had been accustomed since territorial days to hunting unfenced, unposted timberlands at will. Hunters regarded this as a right rather than a privilege.

True to Jones's announcement, Charles Self of International Paper Company and Roy Johnson, O. H. Darling Jr., Ken Clinton, and Terry Hadaway of Georgia-Pacific appeared at the commission's May 1987 meeting. Georgia-Pacific offered to lease 200,000 acres to the commission for $95,000 ($.37.5 an acre) in 1987–88. International Paper offered 28,000 acres for from $.90 to $1.35 an acre. They said the program could be self-supporting through the sale of "leased land" permits. If the commission did not lease this property, they said it would go to clubs and private individuals. Commissioner Tommy Sproles was ready to take the offer, but Commissioner Perry Mikles was more cautious and persuaded his colleagues to postpone acceptance pending reaction at public meetings to be held during the next thirty days. By May, Mikles was ready to accept the proposals, and a divided commission agreed, with Beryl Anthony Sr. of El Dorado and Frank Lyon opposed.

Rocky negotiations followed, principally with Georgia-Pacific. By July, however, Anthony had a prepared statement entered into the commission minutes. It stated,

> I want to take a minute here to thank Georgia-Pacific for their patience in providing the number of acres they have for us and I think they ought to be commended for their actions. This is going to be one of the most important things that we have done or that I have done since I have been on this Commission. This will be the key as to how this works, how the people accept it and treat the land as to whether we get any more or not. This is the picture for the other three big timber companies.

Lyon continued to fight and insisted there should be a fifteen-dollar fee to help cover the leasing costs. There followed a struggle among commissioners about how much hunters would be charged, with the result that the first year was free with ten-dollar permits sold thereafter.

Assistant Director Carl Hunter, who handled the negotiations, said in August 1987 that details had been completed for what had been named the Casey Jones Wildlife Management Area, made up of leased lands under the new program, and that implementation would begin September 1. In September, Director Wilson had this message for Arkansas hunters:

> In recent years, leasing has become a reality. Private timber companies with vast land holdings are looking for new ways to generate income to offset costs in a depressed timber market. The leasing increase has set off a furor in some circles. Hunters of limited income fear they'll lose access to the lands they've hunted for years. They're afraid wealthier hunters who can afford to pay top dollar to lease will create exclusive deer clubs and effectively shut out less economically fortunate deer hunters.

> This is a legitimate concern, and the Game and Fish Commission is aware of this potential problem. So what's to be done? Forbid private timber companies from leasing? Hardly. The ability to privately own land is a foundation of democracy, and within reason, a landowner should be able to do what he wants with his land. (Wilson 1987)

But change was inevitable, Wilson continued, and the commission was attempting to make the best of the situation by leasing and would continue to pursue "opportunities to lease large tracts of private lands" to be kept open to hunters at minimal cost.

Clarence Harris of Sheridan epitomized the hunters who were angered by the timber companies' moves to lease lands to hunting clubs and by a Game and Fish Commission he viewed as letting them get away with it. Harris was a farm-fresh teenager when he served on the aircraft carrier in the South Pacific during World War II where he and his shipmates dodged Japanese kamikaze pilots. He thought he was defending his right to continue living as he had grown up—hunting deer and other game belonging to all Arkansans on vast stretches of land he did not own. There were many veteran hunters in Arkansas who shared Harris's feelings about the timber companies, and they collectively became known as "Gatehaters."

One of Harris's colleagues, Rufus Williams of Prattsville, explained, "We're not saying the property owner doesn't have the right to post his land, but he shouldn't be allowed to lease only to those with big pockets of money" (Griffee, "What the 'Gate Haters' Really Despise Is Change," 1991).

As the so-called Gatehaters fumed, the commission's leased-land program was not going well and appeared to be in jeopardy. In 1990, for example, the commission paid $346,000 to lease three hundred thousand acres from Georgia-Pacific, International Paper Company (Casey Jones Wildlife Management Area), Green Bay Packaging (Cherokee Wildlife Management Area), and a coalition of the Ross Foundation and other Clark County landowners (Big Timber Wildlife Management Area). A hunter could purchase a $10 permit that allowed hunting in season on any of the leased lands for a year. Permit sales brought in $181,256—a loss of $165,144—which meant the program was being subsidized by the sale of hunting licenses and other standard revenue sources. The situation did not improve, and in March 1992, the commission doubled the cost of a permit to $20 to hunt on each of the leased wildlife management areas. This did not solve the problem; however, the landowners balked at lower leasing fees, and the program appeared in jeopardy again in early 1993. A month later, Green Bay Packaging agreed to accept permit fees actually paid by hunters, and by August 1993, new leases had been worked out on the other two wildlife management areas. Finally, the program was on sound footing with leases costing $185,139 and permit sales bringing in $204,680 in the fiscal year that ended June 30, 1995.

As the Game and Fish Commission struggled with the leased-land program, attorney David O. Bowden, representing the Gatehaters, organized the Arkansas Hunting Association and headed by Robert Chumley, filed suit in Pulaski County Chancery Court against the commission, International Paper Company, and other unnamed large corporations dubbed "John Does 500." The commission was accused of delegating wildlife management responsibilities reserved exclusively to it under Amendment 35 to private companies for their enrichment because International

Paper operated a trophy buck program on the wildlife management area the Game and Fish Commission leased from the company. Chancellor Annabelle Clinton Imber admitted from the outset that, "I'm not a hunter—I don't know how the system works." She did know the law, however, and ruled that private landowners had the right to grant or deny access to their property to take game. The Arkansas Supreme Court, she said, had settled this issue as long ago as 1904 in *State v. Mallory*. The court also said, however, this was not an unqualified right because it must yield to the state's ownership of the wildlife and restrictions set to protect the game and fish for "public use."

Chancellor Imber presided over a four-day trial in January 1993 on the question of whether the Game and Fish Commission was violating Amendment 35 by setting more restrictive game-taking rules on some properties such as the Lafayette County Wildlife Management Area (leased from International Paper) than others. After hearing the facts, the judge said the Arkansas Hunting Association had failed to present "clear and convincing" evidence that the Game and Fish Commission had been arbitrary and capricious in its actions. To the contrary, she said, there was sound rationale for every action taken.

Development of the Ouachita and Black Rivers in Louisiana for navigation was authorized in 1871. The project was reauthorized in 1960 to provide for construction of four new locks and dams rather than rehabilitation of old locks and dams. This action came late enough that the change was affected by the environmental movement gathering momentum in the country and in Arkansas, resulting in another federal wildlife refuge—Felsenthal—being purchased to offset adverse environmental impacts. Felsenthal is a sixty-five-thousand-acre tract of bottomland hardwoods interspersed with pine near the junction of the Saline and Ouachita Rivers in Union and Ashley Counties. Its fertile basin is a diverse ecosystem with many kinds of oaks and fruiting vegetation on which deer, wild turkeys, waterfowl, and a wide variety of birds and mammals feed.

Much of this relatively new refuge is (at least until recently) largely unexplored and unvisited—a wilderness harboring alligators, bobcats, river otters, gray and red foxes, squirrels, and raccoons. Large numbers of waterfowl are attracted to the area, and a twenty-one-thousand-acre rest area is included in the development. Access by land was never easy. Much of the refuge is subject to flooding during rainy periods, when water levels quickly rise up to twenty feet.

Sen. Dale Bumpers sponsored legislation that led to Potlatch Corporation, one of the "big four" timber companies in Arkansas, transfering 41,000 acres of wetlands between the Cache River and White River National Wildlife Refuges to the U.S. Fish and Wildlife Service in early 1993 ("Land Swap Will Protect More Arkansas Wetlands," 1993). In return, Potlatch received 17,600 acres of government-owned timberland in scattered tracts in Idaho where the company had a large presence. Part of the 41,000 acres was attached to the White River refuge and part to the Cache River refuge.

Bumpers, who said he wanted the land swap listed at the top of his epitaph as one of his outstanding public service accomplishments, continued to work for appropriations to acquire more Potlatch land to finish linking the two refuges. He also turned his attention to legislation authorizing a swap of some 47,500 acres of mature timberland in the Ouachita National Forest for more than 180,000 acres owned by Weyerhaeuser Company in southwest Arkansas and southeast Oklahoma. About 25,000 acres of the swap were targeted for the new Cossatot National Wildlife Refuge in Arkansas. The proposed trade encountered opposition from traditional Forest Service foes, but the senator saw the bill passed in October 1996.

The Fish and Wildlife Service also bought nine thousand acres of thirteen thousand acres offered by John Hancock Insurance Company along the Little Red River for the new Bald Knob National Wildlife Refuge. Neither the Cossatot nor the Bald Knob refuge could be opened for use because operating funds for the Fish and Wildlife Service have taken a back seat to acquisition monies.

Federal agricultural policies also increased habitat in the 1980s and 1990s, though perhaps only temporarily. For example, to reduce surplus crops and curb erosion, the 1985 Farm Bill created the Conservation Reserve Program (CRP) under which participating farmers agreed not to till highly erodible and other environmentally sensitive lands and to plant them in grasses and trees for contracted periods of time and payments. The program was credited with helping bring quail back in Kansas and producing a bumper crop of migratory waterfowl by 1996. A modified version of CRP was approved through 2002 in the 1996 Farm Bill. As of May 1, 1996, 278,000 acres of former cropland were enrolled in the CRP in Arkansas.

The 1991 Farm Bill established a wetlands restoraton program under which farmers are paid for perpetual or permanent easements on qualifying acreage they nominate to keep out of cultivation. After a slow start, the program accelerated in 1995–96 and more than eleven thousand acres were enrolled in Arkansas by mid-1996. The 1996 Farm Bill reauthorized the program through the year 2002, but added fifteen- and thirty-year easements as avenues to use, giving the program a temporary look.

PART 4

Status of the Species

14 Fish and Fisheries

Fishing is now developed to the point where it is a sport and a recreation rather than a provider of meat on the table. As a recreation, it is only natural that the fisherman should like to pursue the fish in the sportiest fashion.

—W. M. "Bill" Apple, April 13, 1958

The Arkansas Game and Fish Commission operates five fish hatcheries, four warm-water and one cold-water, which make up one of the world's largest state-owned hatchery systems. The agency also has twelve nursery ponds and three net-pen operations. The ponds, hatcheries, and net pens furnish virtually all 10 million to 15.2 million fish the commission uses each year to stock Arkansas's lakes and streams. Net pens are large nylon nets or wire mesh cages suspended in a lake. Catfish are raised in the pens during summer and harvested in fall, and trout are raised in winter and harvested in spring.

The commission also boasts it "has built more acres of lakes than any other state game and fish agency in the country" and that five of the impoundments "are larger than any lake ever built by any other state wildlife agency." There are some 550 boat launch ramps in the state, of which roughly half are provided and maintained by the commission, in addition to scores of fishing piers and walk-in angler access areas.

The Hogan, Centerton, Hulsey, and Donham warm-water hatcheries encompass more than 450 acres of ponds where fish are raised to increase and/or maintain the population of species such as the largemouth bass, striped bass, hybrid striped bass, walleye, saugeye, and tiger muskie. The cold-water facility, Spring River Hatchery, raises trout in raceways and underground silos. Only one other hatchery in the nation uses silos.

It has been said the hatchery system was the only long-lasting accomplishment of the Game and Fish Commission that existed as a department of the state government between 1915 and 1945 (see chapter 12). The ink was barely dry on the Dingell-Johnson Act, which provides federal aid for fisheries work similar to Pittman-Robertson support for game restoration, before fisheries superintendent Joe Hogan was before the commission seeking approval for twenty-five projects. It was

The five fish hatcheries owned by the Arkansas Game and Fish Commission produce millions of fish each year for stocking the state's public fishing waters. At the Andrew Hulsey Hatchery on Lake Hamilton, the first crop of fish was raised in 1941.

Arkansas Game and Fish Commission.

Stocking fish

Seining pond at Hulsey hatchery

Lake Chicot catch

A day's catch of crappie and catfish on Lake Chicot sometime in the 1960s.

Ned Mazzanti.

noted at the time, "These federal aid projects are to be inaugurated under the Dingell-Johnson Act as the first state program under fish management."

Evidence suggests Arkansas's fisheries program, with the exception of a long-running disagreement with fish farmers in the 1950s and an occasional flap about lake drawdowns, has been less controversial on the whole and stronger than its game management efforts throughout the agency's history. Even so, the quality of fishing tends to vary from year to year, depending on climatic conditions; therefore, a long-range view must be taken in order to assess the work of the fisheries division with any objectivity or fairness.

In his March 6, 1955 column in the *Arkansas Democrat*, Bill Apple published a report on Arkansas by the National Sport Fishing Institute based in Washington, D.C. The institute noted that Arkansas had just "passed through its third drought summer in a row [the worst in this series, 1956, was yet to come] and water levels have continued to recede. As a result, a mammoth fish salvage operation was carried out, with many of the commercial food fishes being given to fish farmers for brood stock. Lake and stream beds have dried up and cracked open. The question is will these waters be re-established when normal rains fall, or will they be changed over to agriculture?" The institute reported favorably on these developments in Arkansas:

> Since 1950, the fish conservation picture has begun to change. Less emphasis is on all-out production of fingerling fish to be stocked more or less indiscriminately.
>
> The statewide fisheries survey is now in its fourth year and employs five full-time fishery biologists. Its findings are quickly funneled into the Arkansas-White-Red River Basin studies, and receive daily use by the department. Among its objectives is one to determine the importance and success of stocking.
>
> A fisheries research biologist at Lonoke is studying toxicity to fish life of new insecticides and agricultural chemicals, and the nature and control of fish diseases.
>
> In the Centerton Hatchery, a system of channel catfish propagation has been worked out. Now, the state is furnishing stocks of this desirable food fish to

> lakes and impoundments where there were none present. A commercial catfish fishery involving the propagation of fingerlings for sale has resulted.
>
> Warm-water fish stocks were continued in production for initial and "corrective" stocking are being made, along with winter drawdown, to readjust both fish and aquatic plant life. Needless to say, there is room for readjustment in stocking methods. A step in the right direction was made when two field biologists were located at two hatcheries to review the waters being considered for stocking.

Based on what Bill Apple reported in his wildlife columns in the *Arkansas Democrat*, fishing in Arkansas recovered from the droughts by 1958. Again, however, weather conditions brought about the recovery. On February 9, 1958, Apple said, "Our most productive river, the White, remained at flood stage through the summer. While the excess water in the upper impoundments of Norfork, Bull Shoals and Table Rock was being discharged, the river was bank full. This flood water remaining in the dedicated areas between levees created the finest possible spawning ground. At the end of the season, the White River had more small bass and crappie than at any time during its recorded history."

In his April 26, 1959, column, he said another banner year was in the offing for 1959: "This year promises to be the best in history for the lower White River for it is teeming with bass. The Kentucky bass species has found the water to its liking . . . According to White River commercial fishermen, the bass are so numerous they have reduced the shiner forage to a diminishing point, even going so far as to find the catfish spawn a necessary food."

A decade later, the Fisheries Division reported that its three hatcheries had produced 3,613,361 fish—a decrease in number but an increase in species over the previous year. Catchable-size catfish accounted for nearly one-third of the fish output, producing "instant" fishing in the lakes where they were stocked. This program was succeeding, the division noted, where previous stockings of fingerling catfish had failed. A twenty-acre nursery pond was completed at Lake Greeson in Pike County, the division reported. It likened nursery ponds to refuges producing game such as deer and turkey and said it was a method of establishing a species of fish in a new environment. One of the division's biologists spent the year conducting a survey of the value of commercial fishing to Arkansas. He found commercial fishing, fish farming, and mussel shell industries had a combined value of $12 million to the state.

None of this is meant to suggest the fisheries program has been without problems or that no room has existed for improvement. In fact, a five-member independent study team sponsored by the commission and the National Wildlife Federation made sixteen recommendations for fisheries in January 1970 that included

> A larger portion of fisheries-related revenues should be budgeted for and reinvested in the direct fisheries management and fisheries research function . . . A nationwide study of fisheries expenditures has revealed that the southern states, exclusive of Arkansas, spend an average of 60 percent of their fisheries-related income for direct fisheries management and fisheries research, and that all states nationwide spend a corresponding average of 64 percent, reflecting upward trending allocations of income to these activities.

The team's report shows that in the fiscal year ending June 30, 1968, commission revenues were $4,067,759, of which 52.5 percent came from fishing license

Commercial fish catch, October 1950, White River

Though often controversial, commercial fishing has been an important industry in Arkansas for decades, contributing millions of dollars to the state economy each year.

Arkansas Game and Fish Commission.

sales. Only 11.7 percent of the revenues were allocated directly to fisheries, however, compared to 12.1 percent for game, 23.7 percent for enforcement, and 19.8 percent for administration. In fiscal 1995, 20.0 percent of the agency's budget was allocated to fisheries. Among the team's other recommendations:

> There is an urgent need to provide an adequate staff to carry out the fish management tasks at the district level. At minimum, there should be one professionally-trained fisheries biologist in each district. He should be provided with at least one full time assistant biologist (college-trained), plus necessary stenographic services and adequate summer labor. He should also have access to the hatchery and law enforcement manpower pools for seasonal assistance.
>
> The . . . Division should organize a program of in-service training at the subprofessional level for law enforcement personnel who (under other recommendations elsewhere) should provide cooperative field services to assist in the more routine fish management work.
>
> The practice of supplying free fish for initial stocking of farm ponds, particularly catfish should be stopped unless the recipient pond-owner agrees in writing to permit reasonable public use on request. Otherwise, a reasonable charge should be established for sale of such fish sufficient to defray a reasonable part or all of the cost of production.
>
> A planned program for providing improved physical facilities—i.e., fish disease and nutrition laboratories, fish holding tanks, office space, storage space, residence —is badly needed at the fish cultural stations.

Over time, many of the team's recommendations were carried out.

One of the more interesting episodes in Arkansas's fisheries history occurred in the 1950s when sport fishermen took on fish farmers in a fight the sportsmen could not win because fish farming represented direct rather than indirect economic benefits.

The fight began in earnest in 1956 when Jim Malone Jr., a Lonoke fish farmer and boat dock operator, told the commission and about two hundred sportsmen at a public hearing in January that a recently issued attorney general's opinion gave private property owners the right to raise and sell game fish commercially without restriction or regulation. Malone and H. H. Harvill of Humphrey, spokesman for the Arkansas Fish Farmers Association, said a test suit before the Arkansas Supreme Court might be needed to determine whether the commission could prohibit the raising for sale of game fish or even regulate the business as long as it was conducted on private property and not public waterways. Bill Apple responded at the hearing that the fish farmers were altering their original statements of intent in which they said they would raise only rough fish for resale. He argued that selling game fish would be a tremendous blow to the state's tourist industry that was "tightly tied" to sport fishing.

Apple elaborated on the issue in a January 24, 1956, column in the *Arkansas Democrat:* "Many will wonder why there should be any opposition to what appears to be a new industry for Arkansas," he admitted, adding, "The answer is quite simple. No solution or method yet devised gives proper control over such an operation." He explained,

> Many other states have tried without success to control such operations. A highly profitable bootleg racket is inevitable. Our sister state of Tennessee tried to enforce controls, but only after loud protests from outraged citizens and from adjacent states were changes made in the law. Arkansas and Mississippi found that they were supplying a large portion of the fish to the Tennessee market under the name and guise of Tennessee fish.
>
> If Arkansas permits the sale of game fish, the bootlegger will find a ready market through outlaw operators, and through the back door of many restaurants, and private citizens. The bootleg market unquestionably would undersell and make unprofitable game fish sales of the legitimate dealers.

Netting game fish, 1939

A single drag of the seine netted these men seventeen washtubs full of bass, crappie, pickerel, and other game fish on an unnamed Arkansas stream in 1939. Fearing that outlaw operators would market wild-caught fish and deplete wild stocks, the Game and Fish Commission waged a court battle in 1956 to prohibit fish farmers from marketing game fish raised in their private ponds. A court ruled in favor of the commission, but in 1957, Governor Faubus signed a bill permitting registered fish farmers to raise and sell game fish.

Arkansas Game and Fish Commission.

The pressure and controversy forced the commission in February 1956 to let the courts decide whether it had the authority to forbid the sale of game fish for food that had been produced in private waters by fish farmers. A court ruled in favor of the commission in the summer of 1956, and Apple said in his column (August 26, 1956) he had learned a great deal by sitting through the trial:

> Before the trial, all evidence indicated the fish farmers were interested in raising game fish on a wholesale basis and selling them without restriction.
>
> The trial showed this slant to be incorrect. Everything presented indicated the fish farmers asked only that they be permitted to have enough game fish, principally large-mouth black bass, to control predators in the brood ponds which might jeopardize the stock of rough fish.
>
> The large-mouth bass would act as a control agent, using as its forage food predators which would otherwise prey on the spawn of the rough fish.

> With this condition, the commission has agreed to buy from the fish farmers all surplus bass and place them in Arkansas streams which needed restocking.

The fish farmers were dissatisfied with the court's ruling because, even though they wanted the bass to control predators, they wanted to be able to sell them as well and had legislation introduced in the 1957 General Assembly. The measures would give the Game and Fish Commission authority to register fish farmers and issue permits. The bills, however, contained no provisions for fees to cover the expense of policing the act.

Although sport fishermen were opposed, fish farmers got their law, and on March 24, 1957, Apple reported, "The ink had hardly dried on the bill which Governor Faubus signed permitting the sale of game fish when competition broke out in a big way." Apple continued on April 28, 1957:

> In the last two weeks, one section of the White River gave up between 30 and 40 specially constructed illegal game fish traps.
>
> Such traps are easily constructed of chicken wire and as fast as they are removed, they can be rebuilt. It only stands to reason that the bootleg market will undersell and keep the price of fish below the cost of the legitimate dealer.
>
> While we are complacently simmering on this side of the Mississippi River, friends on the east side have much more to say about our situation with the frank statement "Arkansas has set conservation back 50 years with its passage of an act authorizing the sale of game fish by so-called fish farmers."

The Arkansas Wildlife Federation rallied its opposition in April 1957, but by then, commission director Nelson Cox was defending the agency's decision to authorize fish farmers to put on the market game fish raised (along with long-saleable rough fish) in their private ponds. All fifty-four permits issued to date were temporary, Cox said. If the sale of game fish proved detrimental to sport fishing, he said, as reported in an April 3, 1957, article, the commission would "oppose and prohibit it, granted the constitutional power."

Early anglers on the Spring River

Around the turn of the century, most fishing was done to feed the family or as a commercial endeavor. But Arkansas attracted many visitors who came here to enjoy the sporting aspect of angling. This photo was taken near Hardy.

Arkansas History Commission.

Sport fishermen often find themselves at odds with commercial fishermen, but the two groups found common ground on one issue in the 1950s—outlawing the use of electrical shocking devices for taking fish. They jointly supported a bill in the 1957 legislature that set a mandatory sentence of a year in the penitentiary for persons convicted of using electrical devices. The two groups won this campaign. In his column promoting passage of the bill (July 29, 1956), Apple made some other observations having to do with habitat. He said,

> Unquestionably, the lower White and Arkansas rivers have been the scene of more illegal operations than any other section. The past three years of commercial fishing on the lower White River have been the worst in history, with fewer fish being caught each year.
>
> Of course, this bleak picture can't be entirely blamed on illegal shocking machines. Several factors have affected the fish run, chief of which has been severe drought.
>
> Another big reason has been construction of a levee below Big Creek. This levee in many places was built within 100 yards of the river bank, fencing out much of the overflow water from the bottoms which was normally used by the fish during spawning season. It also curtailed the upstream run of the fish.

Arkansas is freshwater angler's paradise. Natural State fisherman can cast for largemouth on one of the state's many public fishing lakes, try outwitting crappie and bluegills on an oxbow lake, float fish for trout or smallmouth bass on a crystal-clear mountain stream, or drop a line for trophy catfish in a meandering Delta river. Striped bass, spotted bass, redear sunfish, hybrid stripers, walleyes, sauger, pickerels, and a variety of other gamefish also inhabit Arkansas waters.

Bluegill angler

A successful bream fisherman on Jones Lake in the White River National Wildlife Refuge. In Arkansas, bluegills, redear sunfish, and other bream are caught in greater numbers than any other game fish.

Keith Sutton.

BREAM

The Game and Fish Commission now makes no special effort regarding the most-caught fish in Arkansas because it does not have to. In terms of numbers, there is no doubt that anglers catch more bream than any other fish, and chances are excellent a bream is the first fish that a child in Arkansas will catch. Bream occur statewide in nearly every lake, river, pond, creek, or slough and can be caught with everything from cane poles to fly rods. Crickets and red worms are traditional Arkansas bream baits, but a bream will eat almost anything.

"Bream" is a misnomer, however. It actually is a colloquialism under which Arkansans and many other southerners lump a variety of sunfishes. To them, a bream is a bluegill, redear sunfish, green sunfish, longear sunfish, warmouth, or one of several lesser-known species such as the flier, spotted sunfish, and orange-spotted sunfish. The daily creel limit on bream exceeding four inches is fifty a day per angler, and bream are the only game fish legal for bait.

Outdoor writer Charles L. Bridwell says when Arkansans go bream fishing, they usually are seeking bluegills, redears, warmouths, and hybrid bream. Of these, the bluegill is the mostly widely distributed and abundant. Most weigh a half pound or less with an occasional fish up to a pound.

In early 1964, the commission obtained about seven thousand redbreast sunfish from North Carolina and put them in a pond at the Lake Hamilton hatchery near Hot Springs. The fish were only three-quarters of an inch to one inch long when they arrived. In December of that year, about seventy-three hundred mixed adult and intermediate redbreast sunfish were taken from the pond and stocked in the Spring River near Hardy in Sharp County. The fish were of two size groups—five-to six-inch adults from the original stock, and two-to three-inch intermediates, which were the offspring of the imports (Keith 1971).

On January 13, 1966, an additional forty-thousand redbreast sunfish fingerlings were stocked in the Spring River. About fifteen-thousand were stocked in Spring Lake in Yell County on March 4. These release sites were chosen because the redbreast, in its original range, inhabits primarily cool-flowing streams with sandy or rubble-strewn bottoms.

During the winter of 1963–64, the Fisheries Division obtained adult redbreast sunfish from the Virginia Department of Game and Inland Fisheries in exchange for fingerling flathead catfish. These redbreasts were placed at the Hogan hatchery at Lonoke. When harvested in spring 1964, there were eighty-seven adults weighing one-third to one-half pound each.

After this initial effort at stocking Ozark streams with redbreast sunfish to complement the smallmouth bass, the redbreast was never seen or heard of again in Arkansas.

CATFISH

Perhaps the fish most identified with the South in the national eye is the catfish. Entire social events and political fund-raisers are built around feasting on the flesh of the tasty catfish, and fried catfish dinners are found on the menu of any restaurant worthy of the name. Most of these catfish are pond raised because an entire industry grew up around farming them in the 1970s, 1980s, and into the 1990s. There are those who find pond-raised catfish too bland to the taste and will take theirs from the river every time.

Even so, catfish get relatively little respect among sportsmen. As outdoor writer Keith Sutton said in *Arkansas Game and Fish* (September/October 1986), "Catfishing isn't a flashy sport" because it does not involve "fancy boats that could bankrupt a banana republic" and "Volkswagen-sized suitcases full of pretty lures." What catfishing is all about is this, Sutton continued,

> Worms, crawdads, liver and reeking confections of blood and guts are the blue-plate specials of the day for catfish. Trotlines, milk jugs, cheap rods and reels and banged-up johnboats replace the snazzy gear. Glamorous? Hardly.
>
> But don't be misled. When you hook a catfish, you better hold on. They're tough, mean, feisty and smart to boot. Smash 'em, bash 'em, bang 'em up—that's the catfish game.

There are four major members of the catfish family in Arkansas: flathead, blue, channel, and bullhead.

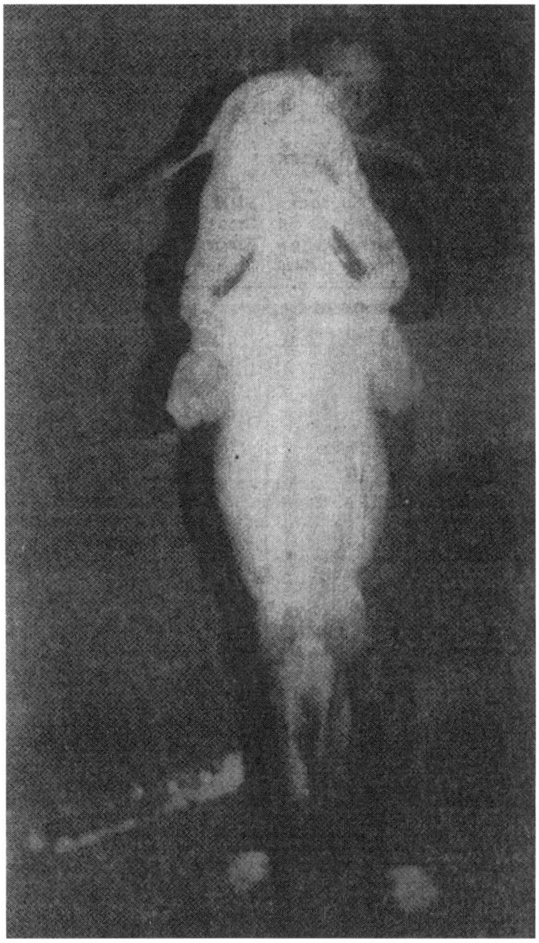

Sayre flathead

Bruce and Mackey Sayre of Little Rock caught a 139-pound, 14-ounce flathead catfish below Terry Lock and Dam on the Arkansas River in 1983. It is the heaviest flathead ever recorded.

The Arkansas Gazette.

The flathead frequents dam tailwaters and large rivers, bayous, and reservoirs. Most average 3 to 10 pounds, but 25 to 50 pounders are not rare, and flatheads up to 140 pounds have been taken in Arkansas waters. The blue rivals the flathead in size and can weigh 150 pounds. It is found in larger streams and reservoirs and, like the channel cat, usually is caught over bottoms of rock, gravel, or sand.

The channel catfish resembles the blue but seldom grows larger than 25 pounds. It is typically a stream fish, proving that catfish will live anywhere and not just in the mosquito-infested, silt-laden bottomland streams of the Delta. They prefer the pools in an otherwise swiftly running river or creek. Channel catfish are produced in the millions at state-owned hatcheries each year and stocked in lakes and ponds.

Bullheads are in virtually all Arkansas waters but hold little interest for fishermen because they are not considered great table fare and seldom exceed three pounds. However, many anglers caught their "bug" to fish as a child by landing a bullhead.

LARGEMOUTH BASS

A sportfishing phenomenon began nationally in the 1970s and escalated in the 1980s and 1990s with the growth of black bass tournaments, often featuring professionals competing for prizes of hundreds of thousands of dollars. Arkansas was no exception in the boom this produced; the number of its anglers easily surpassed one million, counting those who bought licenses and those under sixteen years of age who were not required to have them. The commission encouraged this popularity by marking National Fishing Week beginning in 1986 with an annual Free Fishing Weekend in June during which would-be anglers could fish anywhere in the state with no license required. The commission received five thousand dollars each from Wal-Mart Stores, Plastics Research and Development, and Ensco in 1994 to operate a large portable aquarium that proved to be a popular exhibit at fairs and other public events around the state. The commission also established a Master Angler Awards program in 1985 to recognize those who catch trophy-size fish that are not quite state records but meet demanding minimum weight requirements. To qualify, an angler has to catch a fish meeting or surpassing the weight standard for the species. To be certified as a Master Angler, the fisherman must catch qualifiers in three other categories among eight classes of Arkansas fish—black bass, sunfish, crappie, pikes and true perches, catfish, trout, true basses, and rough fish. A Master Angler receives a hand-letterered certificate suitable for framing and display.

"Those who pursue their favorite game fish helped in this growth, but without question, the major influence on the rapid growth of sportfishing in Arkansas during the '80s was black bass fishing, specifically largemouth bass," commission magazine editor Gregg Patterson wrote. Arkansas hosted five national men's and women's bass championships during the 1980s; many returned in the 1990s. Perhaps the greatest impact on the popularity of black bass fishing in the state came in 1984 and 1985 when the Bass Anglers Sportsmen's Society (B.A.S.S.) held its prestigious B.A.S.S. Masters Classic on the Arkansas River in Pine Bluff. The two events put the national spotlight on Arkansas's excellent bass fishing. Thirty percent of anglers responding to a 1988 commission survey said they fished most frequently for largemouth bass—more than any other fish.

Max Atkinson with bass

Arkansan Atkinson was competing in a national bass tournament, the Red Man All-American, when this photo was taken on the Arkansas River in November 1987. The Red Man and other national events put the spotlight on Arkansas's excellent bass fishing in the 1980s.

Keith Sutton.

Tournaments became so popular the Game and Fish Commission created an information program to track them in 1990. Through it, the commission was able to report in 1992 that 252 bass tournaments were held across the state on twenty-nine water bodies. A total of 6,961 anglers said they fished 56,443 hours and weighed in 12,712 bass of 20,331 pounds. Of the fish weighed in, 87 percent were released. Beaver Lake had the most tournaments with 52, followed by Bull Shoals, the Arkansas River, Lake Dardanelle, Norfork Lake, and DeGray Lake. The commission did not impose catch-and-release on tournaments, "but we promote [them] strongly," Mike Armstrong, assistant chief of the Fisheries Division, said. The tournament tracking program also was targeted to the burgeoning number of small local bass clubs in the state because, Armstrong said, "By and large, these are the groups that need the most improvement."

The news media, both print and broadcast, began reporting tournaments with some of the hype usually reserved for stock car racing and golf. The *Arkansas Democrat-Gazette* sent outdoor editor Steve Bowman far and wide in the country to keep state fishermen abreast of tournaments. He did not have to travel far in 1996 to cover the Operation Bass Red Man All-American Classic, which was held on the Arkansas River at Pine Bluff. After noting Arkansas has more All-American champions and B.A.S.S. Masters Classic champions than any other state, Bowman went on to salute the Arkansas River (June 2, 1996) as the "perfect location" for a national bass tournament, explaining,

> In tournament bass fishing the sport isn't about how many fish are caught or how much weight an angler produces . . . you have to fish the water and conditions that are present, and do it better than anyone else.
>
> That is where the Arkansas River becomes an excellent choice for a major tournament because fishing the Arkansas River is a matter of choices.
>
> There is the dirty, flowing river known locally as the Big Muddy, and there often is clear backwater. Within each of those conditions, there are whole new sets of choices.
>
> There are rocks, wood, jetties and sandbar drops. Then there are cypress trees, laydowns, brush piles, stumps, grass and lily pads.
>
> Each can be fished deep or shallow with crankbaits, worms, jigs, spinnerbaits, Carolina rigs, jerkbaits and topwater lures.
>
> While many fisheries offer much of the same, they often settle into a seasonal pattern that requires an angler to settle on a pattern during a given time.

But on the Arkansas River, those are real choices that must be made every day—and even more so during summer months. Often each choice has to be considered each day and can change in less than an hour.

The largemouth's importance to Arkansas's economy is said to be "staggering" because the state is home to several major lure and bass boat manufacturers, and 80 percent of the country's freshwater baitfish are grown on Arkansas minnow farms.

The popularity caused increased pressure on Arkansas's black bass resource—largemouth, smallmouth, and spotted bass. Numerous questions arose, such as whether tournament fishing caused excessive mortality, adversely affected the spawn, or depleted a lake's bass population. At first, the answer was "yes" to most of these questions throughout the southeastern United States.

Other states began responding to the bass fishing pressure by imposing length and slot limits (limits that forbid anglers from keeping fish within a designated size range, for example a ten to fifteen inch slot limit) on lakes. Arkansas was slow to respond to this, although length limit regulations existed in 1986 on six lakes: Beaver, Bull Shoals, Norfork, Barnett, and Mallard with twelve-inch black bass minimum limits and Lake Maumelle with a fifteen-inch length limit. As the pressure grew on bass, and certain management techniques (such as drawdowns) were proving increasingly controversial, the Fisheries Division began debating the pros and cons of length limits more and more. By November 1986, the division recommended length limits on seven more lakes after public meetings were held in the affected areas. By June 1991, the division "had investigated and proposed management objectives for harvest restrictions, trophy bass lakes, tournament fishing and agency programs and research needs." It offered a statewide largemouth bass management plan the commission adopted as a "guiding document for the implementation of harvest regulations." The plan had the support of Bass Anglers for Arkansas's Future, though its spokesman Vince Miller told the commission, it would have preferred to see a statewide length limit of fourteen inches on all bass species and a creel limit of six.

Aaron Mardis with state record largemouth

Mardis, of Memphis, Tennessee, landed the sixteen-pound, four-ounce record Arkansas largemouth bass in Mallard Lake in 1976.

Arkansas Game and Fish Commission.

A quasi-exotic, the Florida largemouth bass was brought to Arkansas to see how it would fare. In Florida, bass grow much larger than in Arkansas. By accident, Florida largemouths were released into several Arkansas lakes, including Mallard Lake. When a record sixteen-pound, four-ounce largemouth was reported from Mallard Lake in 1976, a hue and cry went up that the fish should not be recognized because it was not a typical Arkansas largemouth and was, without question, a Florida largemouth. The dispute was disallowed, and the record stood (Kaffka 1981).

SMALLMOUTH BASS

A statewide smallmouth bass management plan that was in the works for eight years was adopted by the

Game and Fish Commission in 1995 and went into effect January 1, 1996. It divided the state into two zones: zone 1 covering seventeen counties in Northwest Arkansas and zone 2 for the rest of the state. A minimum length limit of twelve inches and a daily limit of four smallmouths was set in zone 1, whereas the minimum length limit was ten inches with a two-bass-a-day limit in zone 2 with exceptions such as the Mulberry and other extraordinary streams. Fisheries biologist Steve Filipek explained that better nutritional conditions exist in the Ozarks for smallmouths because of the limestone and other factors, which makes them grow faster and larger in that region. Smallmouth bass do not exist in much of the state's eastern and southern lowlands, though they are found in the Ouachita Mountains and Arkansas River Valley.

The smallmouth bass takes center stage in Arkansas's free-flowing streams, but the Kentucky or spotted bass is a strong supporting player. Prolific and adaptable, the spotted is at home in the turbid water of the Arkansas River as well as in clear reservoirs such as Beaver and Ouachita. They also were becoming more abundant by 1996 in such popular smallmouth haunts as the Kings, Caddo, and Ouachita Rivers.

Smallmouth bass angler on the upper Ouachita River, 1989

Smallmouth bass are among the most popular sportfish in Arkansas's cool, clear mountain streams.

Keith Sutton.

Bob Limbird, district fisheries biologist for the Game and Fish Commission, explained the regime for bass:

> Smallmouths are the dominant species in cold water, but Kentuckies are prevalent in areas of warmer average water temperature. Largemouths take over in lower sections of rivers, particularly where they flow into lakes or larger rivers.

Local bass clubs throughout the state often raised money that was donated to the commission for improvements to lakes in their areas. The Arkansas Bass Association got the commission's attention in 1988 when it donated forty-four thousand dollars for the purchase of fish-hauling tanks and an electrofishing boat and to buy Florida-strain largemouth bass broodstock for fishery culture.

Not all bass stories are successes. The Catoosa bass was released in the upper reaches of the Spring River in Sharp County, but it failed to hang on.

ROCK BASS

Then there is a fish called the shadow bass, which is found in only six states, including Arkansas. It is the smallest species of rock bass. Arkansas has one other native rock bass, the Ozark bass. They rarely weigh more than one pound, and most adults are about eight inches long. A Game and Fish Commission fisheries biologist, Sam Barkley of Jonesboro, caught a one-pound, five-ounce shadow bass on the Spring River on October 19, 1994, which was certified by the International Game Fish Association as a world record.

Ozark bass

A catch of Ozark bass—commonly called rock bass, goggle-eyes, or redeyes—from the Buffalo River. The Ozark bass (*Ambloplites constellatus*) was originally described from the Buffalo River at the mouth of Rush Creek in 1977. It was previously thought to be the same fish as the rock bass (*A. rupestris*).

Keith Sutton.

CRAPPIE

Crappie are second only to largemouth bass as the most popular sport fish in Arkansas. Even crappie have felt the pressure of popularity, and in 1993, the commission set the state's first minimum length limits—ten inches or more—on this species at Beaver, Table Rock, Bull Shoals, Norfork, Chicot, and Nimrod Lakes. The daily limit also was set at fifteen on crappie on Beaver, Table Rock, Bull Shoals, and Norfork.

State-record crappie

Richard Fields of Stuttgart displays the former state-record white crappie he caught in 1975. Taken from Akers Reservoir near Ulm, the fish weighed four pounds, three ounces.

Arkansas Game and Fish Commission.

STRIPED BASS

Largemouth bass, crappie, and bream are the bread-and-butter fish found in most man-made lakes in the South. Over the years, the Game and Fish Commission has supplemented these native fish by introducing nonnative species, or exotics.

One of these is the striped bass. Its presence in Arkansas is attributable to Riley N. Donoho of Fort Smith, who was appointed to the commission in 1956. Shortly after he took his seat on the commission, Donoho was visited by a fellow insurance agent. The man told Donoho about a saltwater fish, the striped bass, that became landlocked in South Carolina's freshwater Santee-Cooper reservoirs and was thriving. The stripers were described to Donoho as being terrific fighters, being tasty, and having a potential size of fifty pounds.

Donoho took the message to the commission and received approval to send men and equipment to South Carolina to obtain some stripers. Donoho recalled that Andrew Hulsey, then the fisheries chief, told him stripers were "wonderful fish" that were "just as good as crappie." For all his admiration for the striped bass itself, though, Hulsey also was a bit skeptical because nothing had been done to ascertain what effects, if any, stripers might have on other Arkansas fish. Some of the state's sportsmen felt that stripers preyed on game fish and this made the project controversial.

Still, Donoho and several commission biologists struck out for South Carolina in November 1956. After several relatively unsuccessful nights of trying to obtain stock by fishing in the stump-filled reservoir, professional seiners who knew the lake well were hired to help net the fish. After the fish were caught, they were put into a small earthen pond for holding, but many died from a fungal infection. Of the 350 stripers caught, only 200 made it back to Arkansas, and most of those died before they could be "planted" in Lake Ouachita. Only 27 stripers made it into the lake, but Donoho said much was learned about how to handle and care for the fragile fish.

In April 1957, two trucks were outfitted with water pumps to aerate the water, and an Arkansas entourage was again on its way to South Carolina. On the April trip, fishing for stripers was done below Pinopolis Dam where fish were congregating on their spawning run upriver. Some were netted, while many others were caught on trolling lures. The truck drivers drove day and night to get to Arkansas with 82 fish. Of these, 33 went into Lake Greeson and the remainder into Lake Ouachita (Pledger, "A Commissioner's Dream Come True" 1976). In 1958, 276 striped bass were brought back to Arkansas and put into Lake Ouachita.

These efforts demonstrated, however, that better ways would have to be devised to transport and stock stripers. As early as 1963, South Carolina biologists had been inducing egg production in striped bass by using hormones. Many young fish were flown to the Lonoke hatchery, where they were grown to stockable size.

The stripers led to the development of the nursery-pond system in Arkansas. Nursery ponds, an extension of the hatchery system, are ponds built adjacent a reservoir that are connected to the impoundment by a manually operated gate and drainage canal system. Fish stocked in the pond are raised to desirable size then released directly into the reservoir.

Twelve major reservoirs—Ouachita, Greeson, Dardanelle, Bull Shoals, Norfork, Beaver, Maumelle, Millwood, Blue Mountain, Nimrod, Catherine, and Hamilton—have been stocked with striped bass since 1956. However, the stripers did not reproduce well naturally at first. By 1976, natural reproduction had been found only at Dardanelle and a few other areas of the Arkansas

Stocking striped bass

Striped bass are not native to Arkansas waters. Lake Ouachita received the state's first stocking in 1956 when twenty-seven stripers from South Carolina were released by Game and Fish Commission personnel.

Arkansas Game and Fish Commission.

River. As a result, Arkansas had to import the six million striped bass it stocked between 1956 and 1976, most of them from South Carolina, Louisiana, and Virginia. The fish does not reproduce until six years old.

HYBRID STRIPED BASS

The striped bass quite literally has spawned yet another fish to test the mettle of Arkansas sportsmen—the hybrid striped bass.

Arkansas got into the hybrid striper scene by accident. Stripers were transplanted successfully in Lake Maumelle to the extent that this reservoir was supplying brood stock. Biologists set nets in the upper end of eighty-nine-hundred-acre Maumelle to obtain adult stripers to carry out an artificial spawning program. In 1975, the biologists found many ripe female stripers, but few males. Rather than let the eggs die, the biologists quickly obtained male white bass and fertilized the eggs. Arkansas's hybrid striper program was launched officially later in 1975 when several thousand two-inch fingerlings were released into then-new DeGray Lake near Arkadelphia.

Jerry Shaum with world-record hybrid striped bass

On April 24, 1997, Shaum, of Fairfield Bay, Arkansas, landed the largest hybrid striper ever caught—a twenty-seven-pound, five-ounce fish from Greers Ferry Lake. A cross between male white bass and female striped bass, hybrid stripers were first stocked in Arkansas in 1975.

Joe Mosby.

WHITE BASS

Even during Arkansas's early fishing history, the white bass was rather obscure and did not come into its own until the U.S. Army Corps of Engineers went on its dam-building spree. Then the white bass became a kind of savior of fishing.

The white bass quickly populated the impoundments behind the dams, entering them from feeder streams. After spawning in the tributaries, white bass cruise open lake water as the weather warms in spring. They rise in schools and slash into small shad, making a great surface commotion. It is this preference for feeding on the prolific shad that has made the white bass important to flatwater fishing in Arkansas.

TROUT

It has been noted several times the White River was regarded as Arkansas's most prolific fishery. Even as late as 1950, a seine crew was able to net 346,000 pounds of buffalo in one month from the waterway. This changed when the Corps of Engineers constructed Bull Shoals and Norfork dams.

The most immediately noticeable change was that the cold water released from both Bull Shoals and Norfork dams eliminated the cool-water (principally smallmouth bass) fishery as far downstream as Lock and Dam No. 3 near Mountain View. Severe damage occurred as far down as Lock and Dam No. 1 at Batesville. Much of the food production potential was destroyed by daily fluctuations in water levels that alternately flooded and dried out shoal areas (Mathis 1970).

White bass

White bass are popular panfish caught year-round in Arkansas.

Keith Sutton.

There also were subtle effects that proved even more destructive. For instance, more land was cleared and put into crops when the flood dangers subsided, and the basin's old-river lakes no longer flooded as in the old days, resulting in a decline in fish reproduction. Drainage and channelization programs by various government agencies on tributaries of the White also had severe effects. Because of channelization, the river would rise and fall rapidly after a rain, increasing the silt load. Bottom organisms were smothered, depriving fish of food. Agricultural chemicals began killing fish, especially during the spring spawning season when farmers were using fertilizers and herbicides on their crops.

Earlier efforts by federal and state biologists to stock some of north Arkansas's cooler streams with trout had failed. In 1951, however, the commission decided to try to take advantage of the change wrought in the White River by Bull Shoals and Norfork dams. In cooperation with federal fish hatcheries, the commission began placing fingerlings and catchable-sized trout in the streams. A total of 36,432 trout were placed in the streams the first year. Smaller stockings followed in 1952, 1953, and 1954. In 1956, 64,253 catchable-sized trout were released along with 7,860 fingerlings. The U.S. Fish and Wildlife Service received a $665,600 appropriation in 1956 with which to raise and release six million catchable-sized trout. About $455,000 was earmarked to build a hatchery at Norfork, to be completed in 1957.

The growth rate of trout going into White River and some of its tributaries was considered "phenomenal," but the fish do not reproduce well in Arkansas. Therefore, Arkansas must follow the "put-and-take" system, raising its own brown and rainbow trout—milking the females of their eggs and hatching them artificially.

Wildlife columnist Apple wrote on March 25, 1956, "With three to four million trout being placed in the White and North Fork rivers each year, one only has to close his eyes to picture the fine trout fishing to be enjoyed on the two beautiful rivers." In another column (April 17, 1955), Apple asserted,

> The frigid waters pouring from the foot of these structures [the dams] have created more than 100 miles of river, perfect for trout. It's the only extensive stretch of trout water in the South and one which saves fishermen of Texas, Louisiana, Oklahoma, Mississippi and Tennessee from 500 to 1,000 miles of driving into the Rockies.

Stringer of big rainbow trout

Jim Miller with state record rainbow

In 1984, when organized sport fishermen were fighting the award of a lease by the state Natural Resources Committee to a Mountain View company to take sand and gravel from a bar in the White River, they estimated trout fishing had become a $40 million a year industry.

This was just the beginning, however. The tailwater of the Little Red River below Greers Ferry Dam and that on the Little Missouri River below Lake Greeson proved ideal for trout, as did the cold, spring-fed waters of Spring River. Trout also were stocked in deep, cold-water reservoirs in the state. Arkansas suddenly had a booming business on its hands.

Trout biologist Mark Hudy reported in 1986 that, in spite of an increase in the trout permit price from three to five dollars, a record number of permits were sold, a record number of trout were stocked, and the management objective of stocking rainbow trout greater than eleven inches in all state trout waters was exceeded. Hudy was back again in 1987, telling the commission nearly 120,000 trout permits were sold, and the program showed a net profit of eighty thousand dollars. He said 12 percent of all fishing licenses sold in the state were purchased solely to fish for trout.

By 1990, trout caught from Arkansas waters held seven world line-class records recognized by either the International Game Fish Association or the National Fresh Water Fishing Hall of Fame. Included was Huey Manley's thirty-eight-pound, nine-ounce, all-tackle, world-record brown trout caught from the North Fork River in August 1988. In 1992, Howard "Rip" Collins caught one even bigger in the Little Red River—a forty-pound, four-ounce fish that is the current record holder.

During the 1950s, '60s, and '70s, Arkansas was world-famous for producing large numbers of heavyweight rainbow trout. Miller, of Memphis, caught the nineteen-pound, one-ounce state-record rainbow in the White River in 1981.

Arkansas Game and Fish Commission.

Arkansas waters have produced several all-tackle, world-record brown trout. In 1972 in the White River, Troy Lackey landed a record brown weighing 31 pounds, 8 ounces. In 1988, Huey "Mike" Manley caught the 38-pound, 9-ounce new record trout in the North Fork River. And in 1992, Howard "Rip" Collins caught the current world record, a 40-pound, 4-ounce brown, in the Little Red River.

Ray Glass (left) *and Arkansas Game and Fish Commission* (center and bottom).

The cutthroat trout has been gaining popularity as a trophy catch in Arkansas since eggs of a fine-spotted subspecies were brought into the state in 1983 to be raised in a hatchery and then stocked in the White and North Fork Rivers. Arkansas obtained the eggs in an exchange with Wyoming for smallmouth bass. The original cutthroats were referred to as the Snake River variety after the famous western stream where they are found. The stockings were successful, and the program was expanded in 1989 with fingerlings being released in the Beaver Dam tailwater of the upper White River as well as the Spring and Little Red Rivers.

Cutthroats have taken well to Arkansas's food-rich trout streams. In their home waters, Snake River cutthroats commonly reach twenty inches in length and weigh four to seven pounds. An exceptional fish is as long as two feet and weighs nine pounds.

Above left, **Troy Lackey**

Above right, **Howard "Rip" Collins**

Left, **Mike Manley**

One weighing nearly ten pounds was caught from the White River in 1985, and another estimated at eleven pounds was caught, photographed, and released into the White River by a fly-fisherman in early 1991. There is no evidence cutthroats spawn successfully in Arkansas, and the reason for this mystifies biologists.

Arkansas's trout program was said in 1992 to be a $150 million business. About half the people who trout fish below Bull Shoals Dam traveled more than two hundred miles to get there. Individually, they spent an average of $123 a day, not including fishing tackle and equipment purchases such as boats, motors, and waders. The figures are less on the waters below Norfork ($87 a day), Beaver ($66 a day), and Greers Ferry ($63 a day).

The U.S. Fish and Wildlife Service is the backbone of Arkansas's trout program because its hatcheries at Norfork and Greers Ferry provide more than two million trout stocked annually.

Cold-water releases from Bull Shoals and Norfork that made trout fishing possible in Arkansas also endangered it. Water was released from the bottom of the reservoirs and contained too little dissolved oxygen. Trout need at least four parts per million dissolved oxygen to survive. Fish kills were common because the trout were being smothered.

The Game and Fish Commission began trying at least as early as 1973 to get the Little Rock district of the Corps to change the way it operated Bull Shoals in order to protect the trout stocked downstream. The sticking point was that the Corps must by law operate both dams for flood control and hydroelectric generation. The latter was dictated by the needs of the Tulsa-based Southwestern Power Administration, which had cooperative and municipal customers throughout the region.

Following a 1990 fish kill, the state got the Corps's attention when the Pollution Control and Ecology Department issued an administrative order charging the federal agency with violating Arkansas's water quality standards for the White River, which set a minimum dissolved oxygen content of six parts per million. A classic court confrontation appeared to be in the making; instead, Gov. Bill Clinton's natural resources liaison, Ken Smith, called all the parties to a meeting that led to a solution three years later: modifications costing $138,900 to the electric generators at the dams that added oxygen to the water. No fish kills have been reported since.

WALLEYE

The purchase and distribution of one million walleyes was approved by the Game and Fish Commission on February 20, 1950. The fish were to be released in lakes Norfork and Greeson.

Bill Apple said on September 4, 1955, Arkansas fishermen generally overlooked the walleye in their preference for bass, crappie, and trout. Though thought of as a northern fish, the walleye was "widely distributed" in Arkansas, in the White River near the dam at Batesville, in the Ouachita River, and in Lakes Catherine, Hamilton, and Ouachita. "The Narrows [Lake Greeson] was stocked with walleyes several years ago," Apple recalled, "but few are being caught." How the fish came to be in the Ouachita River is one of those "oral tradition" stories that abound in the South and are all but unverifiable. Supposedly a shipment of fingerling walleyes was enroute

To improve native populations, Arkansas Game and Fish Commission biologists collect walleye eggs and milt each March and raise the resulting fry in hatcheries for later release. In 1982, at the World Walleye Classic in Fairfield Bay, Al Nelson caught the twenty-two-pound, eleven-ounce walleye now recognized as the all-tackle world record by the National Freshwater Fishing Hall of Fame.

Arkansas Game and Fish Commission.

Above left, **Al Nelson with world record walleye**

Above right, **Biologist removing walleye from net**

from a northern hatchery to a western state. The train wrecked near Arkadelphia. Rather than see the fingerlings die, the cleanup crew planted them in the Ouachita and Caddo Rivers, and they migrated upstream well before any of the dams and lakes were built on the Ouachita.

To add variety to the sport fisherman's experience in Arkansas and to save the walleye from extinction following the building of dams, the commission's Fisheries Division regularly produces walleyes for stocking. For example, then Fisheries chief Andrew Hulsey reported in 1967 that 115,000 walleye fingerlings and 5,000 ten-inch northern pike were among the species raised at the Norfork nursery pond and released into the lake. In addition, 400,000 walleye fingerlings were raised in the Bull Shoals nursery pond and released into that huge impoundment.

Greers Ferry was the scene in March 1995 of the seventeenth annual World Walleye Classic. It must have been a success because on May 2, 1996, forty of North America's top professional walleye anglers began fishing the depths of Lake Norfork in the Professional Walleye Trail's first tournament to be held south of Lake Erie. Steve Bowman, the *Arkansas Democrat-Gazette*'s outdoor editor, welcomed the event, telling his readers (May 2, 1996) that

> Traditionally, fishing for walleyes has taken a distant back seat to bass fishing in this state. As a matter of fact, the majority of the walleye caught here are accidents that take place while fishing for bass or crappie. But that could change if these professionals are successful enough to teach us a few things on catching the fish that is considered the No. 1 trophy in the Midwest.

Bowman said the Professional Walleye Trail was aware Arkansas had a walleye record (a twenty-two-pound, eleven-ounce walleye caught by Al Nelson), which was second only to Tennessee's world record of twenty-five pounds. This changed in 1996 when Nelson's fish, caught during the 1982 World Walleye Classic out of Fairfield Bay on Greers Ferry Lake, was named the world-record holder by the Wisconsin-based National Fresh Water Fishing Hall of Fame after the Hall of Fame tossed out the twenty-five-pound Tennessee catch as bogus.

SAUGER AND SAUGEYE

The sauger, distinguished by saddles of dark brown across its back and dark spots on the dorsal fin, is a close cousin to the walleye. The fish is found in the Arkansas, lower White, Little Red, Black, Current, Eleven Point, Spring, and Strawberry Rivers. Anglers are limited to taking six a day.

Arkansas gained another fish in 1987 when saugeyes—a walleye-sauger hybrid—were stocked in Crown Lake in Sharp County after being produced under controlled hatchery conditions by crossing male saugers with female walleyes. Saugeyes are said to be aggressive, good fighters and excellent on the table, and they are good fish for stocking in certain Arkansas lakes because they tolerate a wider temperature range than either walleyes or saugers.

Saugeyes also were stocked in Lake Gillham. Fisheries Division chief Allen Carter explained in 1992 that these two lakes were chosen "because they have relatively low bass populations and because we wanted to see how saugeyes fare in small clearwater lakes like Crown and large clearwater lakes like Gillham. Eating-sized fish are now showing up in both lakes."

Angler with saugers

An angler proudly displays a catch of saugers from below the North Little Rock hydroelectric plant on the Arkansas River. Dam tailwaters on the Arkansas River are the state's most popular sauger fishing hot spots.

Keith Sutton.

NORTHERN PIKE

The exotic northern pike has a sweet tooth for sunfish and was brought into Arkansas to help control the sunfish population in large man-made lakes where they rapidly take over and overrun their territory, resulting in stunted fish. In an effort to overcome this problem, the bream-stocking program in DeGray Lake was limited when the impoundment was new, and the northern pike was brought in. It was thought also that the new fish would provide sport fishing, and it did, for a while, until fishermen forgot about the pike (Kaffka 1981).

RUDD

In 1990, the commission took steps to keep undesirable exotic species from infiltrating the state's sport fishing waters. Some Arkansas fish farmers began cultivating and selling the European

Northern pike

Though no longer found in Arkansas waters, the exotic northern pike was once stocked in DeGray Lake. Pike were never popular with the state's anglers.

Arkansas Game and Fish Commission.

rudd. Dr. Charles Amlaner, the University of Arkansas zoologist on the commission, said the rudd was a dangerous species because it would take over the habitat of native fish and could displace them if released in the wild. Amlaner urged banning the rudd from the state, but staff biologist Mike Gibson said there were only a few rudd dealers because of widespread banning, and he thought the commodity would eventually self-destruct. A regulation was adopted making it unlawful to place the European rudd into any body of water where its ingress to public waters was not blocked entirely. It also was unlawful to offer the rudd for sale to vendors within the state and for a fisherman to use the rudd as a baitfish. Licensed fish farmers could continue to bring the rudd into Arkansas as long as the regulation was followed.

BUFFALO

Arkansas histories are replete with mentions of the buffalo, and many old-timers still fish for them, not minding at all that others overlook these bottom-feeding giants. Three types of buffalo—smallmouth, bigmouth, and black—inhabit Arkansas waters. All feed on zooplankton, insect larvae, and other small water animals, and

Rudd

The European rudd is a popular bait for striped bass and other fish in some states, but its use for this purpose in Arkansas is illegal. Biologists fear the rudd could displace native fish if released in the wild.

Keith Sutton.

Commercial angler with buffalo fish

Buffalo fish are the backbone of Arkansas's commercial fishing industry, with millions of pounds caught and sold each year.

Arkansas Game and Fish Commission.

all occasionally top fifty pounds, particularly in waters with little or no commercial fishing. Buffalo are common statewide in warm rivers and lakes.

BOWFIN

The bowfin, also known as "grinnel," represents the last survivor of a group of fish that was relatively abundant in the geological past in Europe and North America. They are found from Vermont to Florida and the Great Lakes to Texas, being most abundant in the Mississippi Valley. They feed on fish of all sorts and on crayfish. Even small birds sometimes fall prey to them, and their strong sharp-toothed jaws enable them to attack larger fish.

Because of their predatory nature, bowfins are unpopular with most fishermen. Conservationists point out, however, that these predatory fish frequently keep the hordes of suckers, catfish, and minnows in check and thereby help maintain the balance of nature.

The bowfin has gills of unusual structure that may aid them in breathing in foul water. They also can breathe air by taking a gulp at the surface and passing the air into a specialized bladder. At one time, the bowfin "was a much sought after fish merely for the sport of bringing him to the net and throwing him on the bank for the roving wild hogs to consume."

ALLIGATOR GAR

The alligator gar is detested by some sport fishermen who once plied the Game and Fish Commission with pleas to keep the predator under control or perhaps even eliminate it. Bill Apple was among the few who did not hate the gar, and he saw to it that the fish got nationwide publicity. He wrote in one unidentified national publication (a copy of the article was kept in his scrapbook) that

> Deep-sea fishing 400 miles from the nearest saltwater sounds a bit fantastic, but you can get tarpon and tuna thrills on the rivers of southeastern Arkansas.
>
> The "big game" to be found on the lower White River, the St. Francis, Arkansas and Red Rivers, is the huge alligator gar, a saw-toothed denizen of the deep with a fighting heart and the weight to back it up.
>
> The alligator gar, so named because his head resembles . . . that of an alligator, may be found in most southern streams of large sizes . . . He has made himself the popular quarry of two types of sportsmen—fishermen who want something really big on the end of their line, and archery experts who want to test their skill against a tough customer.
>
> Weighing up to 200 pounds and boasting razor-sharp teeth which can cause a man plenty of damage if he should get too close when a gar is being landed, the alligator gar is not an easy prey for either fisherman or archer . . . It takes

deep-sea tackle to land the gar, although some fishermen have done it with lighter tackle after lengthy battles.

Gar weighing up to 200 pounds have been taken from the lower White River. In one 12-mile stretch 600 gar weighing over 100 pounds were taken in one summer season. This was in an area six miles above and six miles below the mouth of Lagrue Bayou. Similar sized big ones have been taken from the St. Francis, Arkansas, Ouachita and Red Rivers.

The sport of using deep-sea tackle to catch the gar, which play havoc with game fish, was originated by a group of Stuttgart residents, among them being Dr. S. A. Drennen, a physician; Dick Deikhoff, a rice planter, and John B. Moore Jr., the latter of nearby Clarendon.

Arkansas' Game and Fish Commission encourages this sport, which is now spreading into northern Louisiana. It helps rid the streams of one of the most devastating predators. Whether you go after him with deep-sea tackle or bow and arrow, the alligator gar is a worthy foe. The sportsman who lands a 100-pounder will know he has been in a battle.

In 1993, the commission established a two-fish daily limit on alligator gar in an attempt to protect the fish.

Man with bowfins

Few Arkansans fish purposely for bowfins, or "grinnel," but many are caught inadvertantly by crappie and bass anglers. Many legends surround these ancient fish. For example, some people say a cooked bowfin will uncook itself if left untouched overnight. Others believe that, given a ritual burial during the proper moon phase, a bowfin will metamorphose into a live snake.

Arkansas Game and Fish Commission.

PADDLEFISH

Another Arkansas fish, the paddlefish (spoonbill), was reported by Keith Sutton in 1987 to be dwindling in number and size because of habitat loss, overharvest, and water pollution. In 1993, the Game and Fish Commission set a new length limit for the paddlefish, replacing one that had been in effect since the 1930s. Under the old

Archer and angler with gator gar

In the lower reaches of the state's great rivers, anglers and archers once battled Arkansas's monster fish, the alligator gar. The number of these giants has dwindled over the years, but occasionally a big one is sighted, bringing back the awe-inspiring excitement of days gone by.

Arkansas Game and Fish Commission.

The paddlefish is perhaps the most unusual fish swimming in Arkansas waters. As commercial fishing for the species increased in the 1980s, Game and Fish Commission biologists implanted radio transmitters in the bellies of paddlefish to learn more about their movements in the state's large rivers.

Missouri Department of Conservation (right) *and Keith Sutton* (below).

Paddlefish

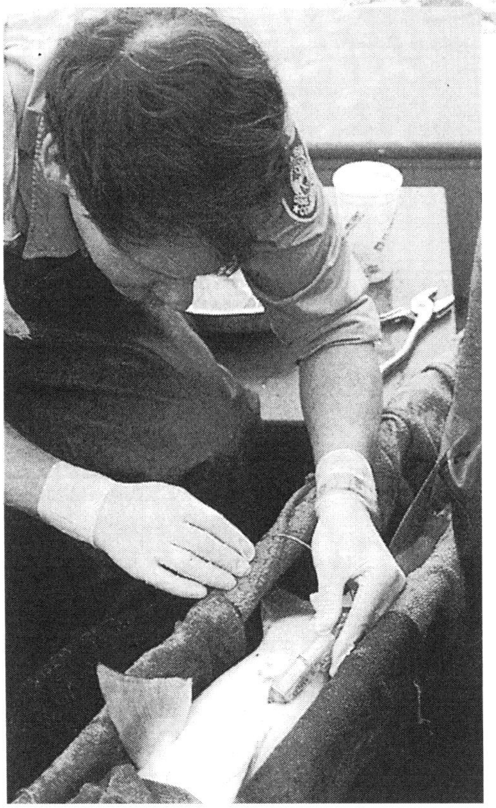

Biologist Steve Filipek implanting radio transmitter in paddlefish

rule, any paddlefish shorter than thirty-two inches had to be released by commercial fishermen. The new rule said any paddlefish under thirty inches from the front of the eye to the fork of the tail had to be released.

Michael Langley of Malvern had a tough job landing the paddlefish that holds the state record in 1994. The paddlefish, which weighed ninety-eight pounds, nine ounces, gave Langley a merry ride in his boat for hours up and down a stretch of the Ouachita River before the drama played out. The previous record for a paddlefish was a mere seventy-seven pounds, four ounces.

Paddlefish were important commercially in the Mississippi Valley in the late 1800s because they were valued for their tasty flesh and their eggs that sold as caviar. About 2.5 million pounds of the fish were harvested in 1899. According to *The Paddlefish: Status, Management and Propagation*, published by the American Fisheries Society, Arkansas's commercial harvest of paddlefish increased nearly 300 percent between 1975–76 and 1981–82 because its eggs are regarded by some to be equal in quality to world-renowned sturgeon caviar from the former Soviet Union. The price of Arkansas caviar was regarded as being impressive, too, ranging from a conservative estimate of $35 a pound to $320 a pound or $20 an ounce.

WHITE AMUR

Arkansas gained national attention as the first state to introduce the white amur, also referred to as the Asian or Chinese grass carp, to the United States. The general backlash, which was not limited to Arkansas, was based on fear of another exotic that had backfired—the German carp, which had been introduced into the United States in the late 1800s (Kaffka 1981).

White amur

White amur, or grass carp, were introduced into the United States in 1963 when the Bureau of Sport Fisheries brought seventy fish from Malaysia to the Fish Farming Experiment Station at Stuttgart, Arkansas. The largest member of the minnow family, the amur is now widely used to control aquatic vegetation.

Arkansas Game and Fish Commission.

Arkansas began experimenting with the white amur in the late 1960s after the Israeli carp proved to be a bust at controlling aquatic vegetation. White amur were placed at the Hogan hatchery because there were severe vegetation problems in many of the state's lakes. Successful artificial reproduction methods were developed over the first few years, and the fish were tried for vegetation control in hatchery ponds with excellent success (Henderson 1978).

Despite objections from other biologists, conservation agencies and even politicians in neighboring states who swore the fish was public enemy number one, the commission released the white amur after controlled experiments had shown it was the most effective weapon in the arsenal for use in the battle with aquatic vegetation (Kaffka 1981).

In 1968, the first lake stocked was Lake Greenlee near Brinkley, an isolated, man-made commission lake encircled with high levees. The fish did control vegetation, and other isolated lakes were stocked with the same results (Henderson 1978).

By 1972, the grass carp was a routine management tool for controlling aquatic vegetation in Arkansas lakes. By 1978, more than one hundred Arkansas lakes with a total of fifty thousand acres had been stocked with grass carp (Henderson 1978). Grass carp raised in Arkansas are hybrids that do not normally reproduce in the wild. Missouri outlawed the fish when Arkansas prompted the national furor by stocking it, but six years later, the Show-Me State rescinded the ban, having discovered the fish was a "paper tiger" (Kaffka 1981).

After the fish reached a suitable size and upon controlling vegetation in Arkansas lakes, impoundments were opened to commercial fishing in order to remove grass carp. This met with varied success (Henderson 1978). On a statewide basis, more than twenty-five tons of grass carp were taken in 1976 by commercial fishermen. Yet the grass carp made up less than one percent of the 6.5 million pound commercial catch of 1976. In 1995, Horseshoe Lake in Crittenden County yielded a white amur weighing sixty-five pounds, fourteen ounces, surpassing the world mark of sixty-four pounds caught the year before on Little Blakeley Creek off Lake Hamilton.

CRAYFISH

Easy to overlook but important in a healthy ecosystem is the crayfish, which Arkansans refer to as "crawfish," "mudbug," or "crawdad." More than five hundred species occur worldwide, about half in North America. Exactly how many species are present in Arkansas is unknown, but knowledgeable biologists believe there are more than forty species and subspecies. The endangered Hell Creek Cave blind crayfish of the Ozarks is white. Other species are tan, red, green, brown, black, or a mixture of colors. They vary in size from three-quarters of an inch to the six-inch red swamp crayfish that is the star of Cajun cooking.

Biologists say it is virtually impossible to find a moist place in Arkansas that is not crayfish habitat. (In fact, at least one Arkansas crayfish lives in seemingly dry yards and pastures where they dig burrows down to moisture—perhaps twenty to twenty-five feet.) The largest crayfish populations are in shallow, fertile, relatively warm waters with mud bottoms and plenty of vegetation.

Crayfish are considered an important part of the environment because they are part of nature's clean-up crew as they convert dead and decaying plant and animal matter to useful protein once again. They provide food for many fish, mammals, and birds and are a popular and effective bait for bass, catfish, and trout. Dermott in Chicot County annually salutes the invertebrate with a festival. Outdoor writer Jim Spencer declares that, regarding eating crayfish, "Don't knock 'em if you haven't tried 'em."

Crayfish

The increasing popularity of boiled crawfish, crawfish étouffée, and other dishes has led some south Arkansas farmers to raise crayfish as a second crop rotated with rice.

Keith Sutton.

15 Game and Nongame Species and Habitat

The people of Arkansas had spoken with the adoption of Amendment 35. Their message: Get serious about the state's wildlife.

Actually, there is nothing in the record to suggest the Game and Fish Commission was anything but serious about restoring Arkansas's wildlife. It was hamstrung, however, by politics—that is, by a lack of authority to do what was needed. Also there was little research on the factors contributing to the decrease or increase of game populations. Writing in *Arkansas Game and Fish* (fall 1976), the late Harold E. Alexander, one of Arkansas's most honored conservationists, put it this way:

> Wildlife management, as we interpret it, really began about the time of the establishment of the wildlife research units some 25 years ago. There were, of course, numerous practitioners before that time, but wildlife management as a profession developed with the establishment of a positive system of education to train biologists and managers. Since then, we have in some respects come a long way. In others, we have failed or floundered about, without accomplishing our purposes . . .

The case of the pheasant exemplifies how a lack of knowledge hampers even well-intentioned efforts. In the 1920s, the commission tried to reestablish pheasants in Arkansas through the game farm in Benton County. For the next fifty years, several species of pheasants were introduced, and the results were the same—"not the success hoped for." It now is known that pheasants will not adapt to Arkansas for biological reasons. The moral here is that even when there is habitat, "the picture is much more complex and there are all kinds of biological impacts to be considered."

In the early 1940s, before Amendment 35 was approved by the voters, the commission concentrated on three popular game species: the northern bobwhite, the wild turkey, and the white-tailed deer.

Quail hunter and dogs

This hunter sought his prey near the Pope County Quail Area in January 1942. Bobwhite quail have always ranked high on the list of Arkansas's most popular game species.

Arkansas Game and Fish Commission.

NORTHERN BOBWHITE

The demonstration projects the commission established in the early 1940s were described in part 2. Considerable success was attained in almost all areas, and surplus birds were trapped and transplanted. The quail population declined in a few areas, but this usually was attributable to the change from general farming to cattle grazing. Since that time, experiments have shown that native quail can recover sufficiently where there is suitable habitat and where sufficient brood stock exists.

Beginning in 1942, the commission directed almost all of its quail restoration efforts toward establishing field border strips. The strips consisted of a belt of *Lespedeza bicolor* (shrub lespedeza) adjacent to woods, paralleled by a strip of sericea lespedeza next to a cultivated field. Seedlings were planted in quail food plots at Blue Mountain, Nimrod, Booneville, and Camp Robinson and in turkey food plots on the Norfork and Shumaker areas. "All of these areas are publicly owned, and thus far, we have not had some of the disappointments which we have encountered in attempting to restore quail on land belonging to private individuals," the commission was told in 1950.

Following passage of Amendment 35, the commission continued and expanded the quail program. In February 1948, it initiated a statewide game management contest "designed primarily for more intensive management on quail and small game production." Each entrant was required to sign up for game management development on at least 160 but not more than 460 acres considered suitable for wildlife habitat.

Commission staff member Wayne Sloan told Director McAmis in a memorandum dated March 18, 1950, that all counties in the 1949 quail restoration contest had been judged and the prizes awarded except in Craighead, Crittenden, Lee, Phillips, St. Francis, Woodruff, and Baxter Counties. "We are finding better plots than we expected, although we did not have enough good plots last year," Sloan said. At that point, he added, there were 5,957 applications submitted in the 1950 contest. Later that year, Sloan said he felt sure there would be "3,000 good plots,

1,000 fair plots, and about 1,000 only partly complete out of the 6,000 original applicants." Food and cover, he said, were "very good this year. The late hatch was good and I think we will have a few more quail this year than last. Overall conditions are good." Sloan also reported the 1951 contest would have a minimum of 10,000 contestants. Although this would indicate a growing interest in the contest, its popularity waned, and it was discontinued in the mid-1950s because it became an administrative burden.

Habitat appears to be the key to quail populations, though weather also accounts for yearly fluctuations in the number of birds. Bill Apple knew the interlocking relationships in habitat, weather, hunting, and other factors in quail populations. He noted in his column on January 24, 1954, that "the past two seasons have been rugged ones for the quail, two drought years combined with natural predators, reduced the stock to the point that the kill was extremely light, and the critical situation moved the . . . Commission to closing the season early."

He advised sportsmen's clubs that they would get greater returns if they abandoned efforts to pen-raise quail and concentrated on increasing quail food and cover. Even so, he warned, the number of coveys on a given area is limited, the maximum point being about one quail per acre because the birds do not tolerate overcrowding. Apple stated bluntly (January 24, 1954):

> While restocking is advocated by many sportsmen, it is not the final answer. According to authoritative estimates, we have over four million quail in the state at present, and due to the small percentage of survival, the high cost of pen-raised birds, the expense of releasing, means that the limited number of birds that could survive would be such a small percentage of our present total crop, that it seems folly to pursue that course.
>
> The answer lies with the organized sportsmen's clubs . . . who are willing to devote time, money and personal effort to the problem, and seek the full cooperation of the landowner.

Ten months after Apple wrote this, the quail picture had brightened considerably, with at least fifty counties reporting more quail than at the same time the

Inspecting quail food plot

In July 1970, officials of the Arkansas Game and Fish Commission inspected a quail food plot on Galla Creek Wildlife Management Area. *From left,* Troy Whitlow, area manager; Thomas Taylor, biologist; and Randall Taylor, district biologist.

USDA Soil Conservation Service.

previous year. "The plucky game bird is making a somewhat miraculous comeback," Apple commented on October 24, 1954. His information was based on a late quail census and whistle count by rural mail carriers covering their routes. By mid-August 1955, quail indeed had staged a recovery: "Favorable spring weather has brought one of the finest hatches of quail we have had in many seasons. The pairs unfortunate in losing their first hatch reestablished their nests with a successful new brood . . . all signs point to a bumper crop of quail as compared to last year. Where one covey was present last year, two to five coveys are now being found," Apple said.

On December 8, 1957, Apple wrote that sportsmen who had been "wagging their heads sadly when the subject of quail hunting is mentioned, longing for the good old days when coveys were plentiful and large . . . are caught short. For the quail are plentiful again . . . In my estimation, there are more quail this year than in the past 20 years." There were several reasons the quail "made such an astounding comeback," he continued. "Favorable weather conditions have provided good cover and food, and the Soil Bank program has taken much land out of cultivation—leaving it for the production of food and cover for birds."

The 1985, 1990, and 1996 federal farm laws contained a Conservation Reserve Program under which farmers were compensated for taking highly erodible and other ecologically sensitive acres out of crop production and planting them in grasses or trees. Some hoped this program would achieve the same success, in terms of improved quail habitat in the 1980s and 1990s, that the Soil Bank had in the late 1950s. Because most lands were planted to sod grasses and pine trees and because farmers were required to mow fields, it apparently did not do this for quail.

The Game and Fish Commission knew 73 percent of Arkansas's license-buying hunters hunted deer while only 17.5 percent hunted quail. This translated into about thirty-six thousand hunters who pursued quail during the 1984–85 season. Private industrial lands were hunted most often, followed closely by public lands and private nonindustrial lands. The average hunter was thirty-six years old and bagged twenty-two quail during the season. Quail hunters constituted the state's fifth largest hunting group. To hear commission members complain at their meetings about the paucity of quail, one could only assume each had been contacted by all thirty-six thousand of the bobwhite hunters. Yet the following, which appeared in a 1987 issue of its magazine, summed up what the commission knew to be the problem and the challenge it faced:

> Wildlife management efforts can be summed up in three words: habitat, habitat, habitat . . . The battle to increase Arkansas' bobwhite quail population will be won or lost through habitat manipulation.

Ninety-one percent of the land in Arkansas is privately owned, so the commission had long realized that if habitat was to be reclaimed for the bobwhite, it would have to be done largely through the efforts of private landowners. Several articles appeared in the agency's magazine during the 1980s, telling landowners in detail what they could to do improve conditions for quail, rabbits, and other small game. For example, fescue, planted heavily by stockmen in northwest Arkansas in the 1940s, was discovered to be a habitat problem for quail. It provides little if any food and is grazed and cut so low that it provides no cover. The commission began encouraging farmers and ranchers to convert to a mix of warm and cold-weather grasses on their lands.

Though not specifically designed for quail, an outreach program called Acres for Wildlife, operated by the commission in conjunction with the University of Arkansas Cooperative Extension Service, improves habitat for wildlife on private lands. To participate, a cooperator must agree to maintain, improve or create wildlife habitat on lands totaling at least one acre for at least one year. The commission provides participants with seed packets to produce food plots attractive to game and, at one time, provided bundles containing twenty-five tree and shrub seedlings that produce nuts or fruit useful to bobwhites and other wildlife.

In 1991, staff wildlife biologist Jane Rice was assigned the challenge of developing a plan to increase quail populations through intensive habitat management on selected areas. Two additional biologists were needed to carry out the plan, and the commission adopted an order in which it authorized an additional $100,000 appropriation. Two additional persons were hired, but assigned to general duties rather than to the quail management program.

Game and Fish Commission magazine writer-editor Keith Sutton conceded in a November 1989 issue of *Arkansas Game and Fish* that the glory days of quail hunting in Arkansas were over but said the sport still attracted numerous devotees and that good hunting could still be found statewide "if you take time to seek it out." He advised,

> Bobwhites are rarely found in the middle of a cropfield or in heavy woods. They favor edge cover, where two types of habitat intersect. This might be a brushy fencerow running through a cropfield, a briar patch at a pasture's edge or brushpiles adjacent to a small woodlot. Look for roosting areas like heavy brush and native grass fields, escape shelter like hardwood thickets and shrubby property lines, and feeding areas with a variety of grains, wild seeds, legumes, forbs and insects. Finding appropriate cover with plentiful edge is the key to locating quail.

WILD TURKEY

When Arkansas conducted its federally financed game survey in the 1940s, there were about seven thousand wild turkeys in the state distributed over forty-seven of its seventy-five counties. The total reported kill was 202 in 1938, 299 in 1945, and 145 in 1949 (Holder 1951). A limited restocking program had been in progress since 1932. Some were native turkeys trapped and relocated, and others were pen-raised and released. Only 300 turkeys were killed in Arkansas during the 1946 spring season—1 more than in 1945.

This prompted a commission staff member to recommend a "serious study of the overall turkey situation" because "drastic measures must be taken this year in order to save the turkey as a game bird in our state." The turkey picture became even more grim the following October when the commission learned that the $8,074 it had appropriated for turkey restoration could not be spent: federal funds had been reduced, and available monies had been allocated to other projects. Four months later, however, federal aid coordinator Trusten Holder received commission approval to establish a turkey project on the Indian Head area west of Lake Norfork, so the turkey effort was not abandoned entirely.

The turkey situation was so critical by February 1948 that the commission voted a statewide closed season. A year later, however, it approved a five-day season with a one-gobbler bag limit. In March 1950, an emergency closing order was issued for

In the 1950s, efforts to stock semitame, pen-raised turkeys in Arkansas proved unsatisfactory. Trapping wild birds with cannon nets and releasing them in low-population areas proved to be a more productive method of restocking.

Arkansas Game and Fish Commission.

Turkey being released from truck

Removing turkey from cannon net

counties bordering the Mississippi River because turkeys were being disturbed by a substantial overflow. Johnson County also was closed because it had just received an initial restocking of turkeys.

In the 1950s, the commission had considerable success restoring deer by trapping them in areas of high concentration and restocking them in areas with few or no deer. It long had been thought that a similar program would work for turkeys, and the commission decided to begin an extensive turkey restocking program.

Commissioner R. N. Donoho of Fort Smith tried all the approaches he could think of to boost the turkey population. While attending an international meeting of game and fish commission officials in 1956, he became acquainted with one of his Pennsylvania counterparts. Donoho was not a successful insurance company owner for nothing; he managed to talk the Pennsylvanian into donating some turkey eggs to Arkansas. They were hatched and released in Arkansas, but their fate was not documented.

Stocking efforts with semitame birds proved unsatisfactory. Hunters usually found them feeding at a farm yard instead of in dense woods, and shooting a turkey perched on a pump handle near the watering trough just did not have much sport to it. Moving turkeys from some parts of the state to others began to pay off, however, because the season kill in the late 1950s rose to 566 (Kaffka 1979).

In 1960, the commission leased eight-thousand-acre Brandywine Island from the U.S. Gypsum Company, a pulp manufacturer. The island is in the Mississippi River north of West Memphis in Crittenden County. Its large turkey population furnished an average of 150 birds annually for restocking elsewhere in the state. White River and Wapanocca National Wildlife Refuges and St. Francis National Forest also provided some of the turkeys used in restocking through March 1970.

Turkey trapping usually starts in January and runs through March 15. Trapping sites are baited as much as a month in advance. A net measuring thirty feet by forty feet is fired over feeding birds by three small cannons triggered electrically from a blind some distance away. The captured turkeys are banded. The birds then are placed in cardboard delivery crates, shipped to their new location, and released, preferably on the same day they were trapped to avoid prolonged confinement. Trapping mortality is about 8 percent (Gaines 1972).

The minimum number for a successful release is sixteen turkeys, with a ratio of six gobblers to ten hens. Three or four years provide the released turkeys sufficient time to reproduce and increase their numbers, after which the population can support limited hunting. However, if there are few or no signs of increase within four years, "efforts to restock . . . should be abandoned until the limiting factor is determined and corrected." More often than not, the limiting factor—at least in Arkansas—is man and his disturbances, particularly during turkey nesting seasons because the birds frequently abandon their nests if disturbed.

By 1972, there were an estimated twenty-thousand turkeys in Arkansas. Winona Wildlife Management Area west of Little Rock at one time had only one flock of turkeys remaining. Now it is a prime hunting area. The success has been attributed to the use of a different stocking approach. At first, only adult birds were stocked in the fall. Then the commission began stocking a hen and her brood at Winona in the spring, adding older birds in the fall. Biologists found the hen and the brood would stay in the area where stocked, whereas the older birds would travel up to ten miles from their release point (Rush 1973).

The turkey restoration project was costly. In 1972, each trapped bird represented a cost of $266, with the commission being reimbursed for 75 percent of it by the federal government through excise taxes on sporting equipment under the Pittman-Robertson Act (Gaines 1972).

In 1962, when the Game and Fish Commission began to take a more scientific approach to game management, the first turkey season was held in seventeen counties. By 1965, the number of counties grew to forty-five and in 1964, the harvest topped one thousand for the first time (Kaffka 1979). Since then, the annual harvest has increased steadily, and hunters are bagging more than nine thousand birds each year. Most of the harvest occurs during spring.

Sharp of eye and extremely wary, the turkey presents a unique set of challenges to the hunter, and that is why Apple, in a March 27, 1955, article, called it an exciting sport:

> We might go so far as to say that once a fellow gets a taste of turkey-hunting, he'll not give up on pursuing this crafty and elusive bird until the very sundown of life. In fact, I personally know a number of fellows who are this day as avid and ardent in the sport as they were 10 or 15 years ago. And some of them have yet to bag a gobbler.

Wildlife scientists attribute the comeback of turkeys in Arkansas and elsewhere to careful protection, restricted hunting seasons, and successful efforts to trap and move wild birds to areas

Turkey hunter silhouette

A successful fall turkey hunter in the Ouachita National Forest, Montgomery County, 1989. The 1997–98 fall turkey season was closed to guns and restricted to bearded turkeys.

Keith Sutton.

where there were none. By 1986, Arkansas had trapped and relocated more than 5,000 turkeys and was continuing the program with roughly 300 birds a year, most of them going from the Ouachita Mountains and areas along the Mississippi River to the Ozark Mountains and the Gulf Coastal Plain. The estimated 1986 statewide turkey population was 100,000 with enough good turkey habitat statewide to support an additional 100,000 to 150,000 thousand birds. The reported harvest that year was 6,599 in spring and 928 in fall. More than thirty thousand hunters were said to take to the woods in spring to pursue turkeys (making it Arkansas's seventh most popular hunting experience), and twenty thousand participated in the fall season in 1986. By the mid-1990s, turkey hunting was reported to be the fastest growing of all the permitted sports. Recreational survey data indicated Arkansas turkey hunters each spent at least $275 a year on goods and services related directly to their sport.

Spoiled by success, Arkansas turkey hunters were alarmed by poor seasons in the late 1980s. Some also were aware that, despite the fact that Missouri had started its restocking program at the same time Arkansas did, hunters in the "Show Me State," with a shorter season, were bagging about fifty thousand turkeys annually, even with a shorter season. The Arkansas commission responded by setting a shorter spring season on the recommendation of staff biologist Ron Smith, who had concluded that the long season was affecting the state's turkey population adversely. He explained,

> A long season increases the percentage of juvenile birds in the kill. Last spring, our kill was 35 percent jakes. That's too high, and it affects the number of 2-year-old gobblers available to hunters this year.

Smith said a later-opening season provided needed time for mature gobblers to mate before hunters removed them. The shorter season was aimed at reducing the percentage of juvenile gobblers in the harvest. He continued:

> Sixty-five percent of the kill in past years has come in the first 10 days of the season. If our objective was to reduce the harvest, we'd need to shorten the season to less than 10 days. That's not our goal. What we're trying to do is alter the structure of the turkey population, so the quality of the spring season doesn't depend so heavily on last year's hatch. To do that, we have to reduce the kill of juveniles, so they can carry over to become adults.

The question of restocking Arkansas with birds from other states was raised in late 1990 because of a population downturn in the Ouachita and Ozark Mountains. Staff biologist Gregg Mathis said research was needed on habitat conditions before implementing an intensive restocking program. This prompted the commission to direct the Wildlife Management Division "to take an immediate look at the turkey population statewide with intentions to restock as soon as possible."

In the interim, staff biologist Mike Widner reported that the 1991 spring turkey harvest set a record with an overall 29 percent increase over 1990. The total reported 1991 harvest was 8,367. Two months later, staff biologist Dr. Joe Clark came to the commission with a four-year research proposal, "Population Biology of Wild Turkeys in Arkansas," to be conducted on the Ozarks' White Rock Wildlife Management Area. The proposal encountered immediate objections from Commissioner Perry Mikles of Booneville, who wanted the study done in the Ouachitas. Dr. Clark countered that it would be a "few years before research financing" could be obtained for that area. The study was to be extensive, involving trapping, examining, taking blood samples, leg banding, and following movement through radio transmissions on the turkeys in the Piney Creeks and White Rock Wildlife Management Areas with the objective of learning more about mortality, reproduction, hunting's effects on population, habitat for successful nesting, genetic differences, and improving management techniques. The commission approved the study with the provision, initiated by Commissioner Hal Hunnicutt of Conway, that Dr. Clark come back with a similar proposal for the Ouachitas in sixty days and implement it "within the next 12 months." The study was funded in part over the first three years by $75,000 from the National Wild Turkey Federation.

A similar study began for the Ouachitas in 1993. Meanwhile, the spring turkey harvest in Arkansas reached 9,624 in 1992 but dropped to 8,700 in spring 1993. Something was working because a record 10,422 turkeys were reported harvested in spring 1995 in seventy-two of the state's seventy-five counties. Even Faulkner County, where the birds were so scarce that hunting had been closed until the early 1990s, had a reported kill of 106.

Restocking continued. For example, in April 1995, sixty-four turkeys were trapped in areas where they were plentiful and relocated to Benton, Lonoke, Woodruff, and Cross Counties. Seven even went to neighboring Louisiana as part of the National Wild Turkey Federation's Project 2000.

Reported Wild Turkey Harvests*

YEAR	TOTAL	YEAR	TOTAL	YEAR	TOTAL	YEAR	TOTAL
1940	153	1955	261	1970	1,164	1985	5,697
1941	305	1956	313	1971	1,952	1986†	7,527
1942	365	1957	461	1972	2,231	1987	10,792
1943	364	1958	461	1973	2,495	1988	9,326
1944	318	1959	317	1974	3,211	1989	9,421
1945	299	1960	566	1975	4,083	1990	7,146
1946	Closed	1961	627	1976	4,264	1991	8,682
1947	Closed	1962	711	1977	5,711	1992	9,917
1948	Closed	1963	768	1978	6,266	1993	9,063
1949	195	1964	1,229	1979	5,874	1994	9,155
1950	145	1965	1,581	1980	6,704	1995	10,684
1951	201	1966	1,634	1981	4,096	1996‡	9,438
1952	363	1967	1,455	1982	4,369	1997‡	10,692
1953	350	1968	1,634	1983	5,356		
1954	303	1969	1,129	1984	5,913		

*Combined spring and fall seasons.

†Game and Fish Commission began reporting turkey harvests on a fiscal year basis; to maintain consistency, this table continues to represent turkey harvests on a calendar year basis.

‡Spring season only.

Source: Wildlife Management Division, Arkansas Game and Fish Commission, Little Rock, Ark., 1996.

WHITE-TAILED DEER

In the annals of North American wildlife management, few success stories equal that of the white-tailed deer. Some contend that the impressive whitetail record is a direct result of scientific management. Others point to the animal's innate resilience to altered environments, particularly those of human design. Conversely, there are those who consider the recent history of white-tailed deer anything but a success. They are quick to note crop and other property damages and highway accidents that are direct consequences of abundant whitetail populations. These skeptics argue that the whitetail has received too much investment, prompted by the demand for recreational hunting, from the wildlife profession and that attention to other species and biological problems have suffered as a result. These differing perspectives will bear significantly on future attitudes toward white-tailed deer and their management in some areas of the country. However, no one can deny that the whitetail's modern history has been remarkable.

The whitetail's recovery nationwide is reflected in Arkansas. As noted in part 2, the state's deer population fell to an estimated five hundred in 1930 (Holder 1951). Today, the population is reported to be at least one million deer.

Legal deer harvest numbers grew steadily and began soaring in Arkansas in the late 1980s and 1990s, reaching a record 163,924 in the 1995 season. Game and Fish Commission director Steve N. Wilson cited several factors for the record kill: the issuing of an antlerless bonus deer tag to hunters for use in southern Arkansas, zone quota permits for antlerless deer in northern Arkansas, more habitat as a result of the Conservation Reserve Program, and a "growing group of younger hunters who haven't been instilled with a bucks-only" mentality.

When the Amendment 35 Game and Fish Commission took office in 1945, it found a fairly extensive deer stocking program in progress. In 1942, the old commission purchased 107 deer from the Sandhill Game Farm in Babcock, Wisconsin. An additional 96 were purchased in 1943, and 101 were obtained the next year (Holder 1951). Then good fortune befell the agency. On June 19, 1944, the old

Deer hunter and boys with doe

The settlement of Arkansas placed considerable stress on whitetail populations and their habitats. Habitat loss, subsistence use, and marketing of hides and venison contributed to the near extirpation of the species. When this photo was taken in 1921, the taking of even a single deer was an uncommon event.

Arkansas Game and Fish Commission.

commission adopted member Ashley Johnson's motion to buy a truck or jeep for a deer restocking program. The move was fortuitous because, four months later, assistant commission secretary W. N. Deaton reported on a meeting he had had with Ozark National Forest officials and on an inspection of the Sylamore District. There was conclusive evidence, he said, of an overpopulation in the Sylamore deer herd, and forest service officials felt drastic measures should be taken to reduce it. They suggested establishing an open shooting season in the refuge area, opening the season on does, or even permitting the use of dogs. Deaton revealed that he had taken a "firm stand" against any of these measures. The commission supported Deaton and voted to expand its limited trapping program in order to get as many Sylamore deer as possible moved to its refuges. By January, Deaton reported that 92 deer had been trapped at Sylamore and released in state refuges. The staff, he added, planned to trap a minimum of 200 during the season. By February, the number of transferred deer had climbed to 142, and by April, the staff had exceeded its goal with 262. By August, the total had mushroomed to 318.

The stocking was even more successful in 1946 when the total reached 516 with a 7 to 1 ratio of does to bucks. Of the transferred deer, 28 came from the Howard County deer refuge and 488 from Sylamore. Further, the statewide deer population was reported to be 31,597, and the animals had been established successfully in sixty-six of Arkansas's seventy-five counties (Holder 1951). Deaton said he felt that "within a few years, deer hunting will take its place as one of the major attractions during the fall hunting season. Last year, the wardens reported checking a total of approximately 14,000 deer hunters in the field during the two periods."

A crisis developed at Sylamore in 1949 when hunters swarmed into the district, seemingly bent on slaughtering the deer herd. The commission directed McAmis to enlist the aid of federal government agencies, including the FBI, as well as the state police "in order to clear up the situation." He also was told to assign additional wardens to the area and "have them patrol as long as they were needed."

A month later, McAmis reported 103 deer had been trapped in one month at Sylamore and had been released in Izard, Fulton, Newton, Perry, Sebastian, and Hempstead Counties. The commission terminated the deer trapping program, effective April 24. Later at the same meeting, the commission closed Lonoke County and parts of Prairie, Pike, and Boone Counties to deer hunting for five years and recommended that additional deer be released in Jackson, Union, Prairie, and Boone Counties and in the Devil's Den area. Five months later, the commission asked federal aid coordinator Trusten Holder to "make an immediate survey of the [deer] trapping program so as to fully advise the commission on the plans and needs for trapping during the coming season and also to investigate fully the deer population in the area to be trapped."

That the deer restoration program was well on its way to success was evident in a report new assistant commission secretary Lewis Polk and Holder gave in September 1950 on the areas closed to deer hunting five years before as a protective measure for newly stocked animals. They were eligible now to be open to hunting and should be opened, the staff said. The commission agreed and opened Baxter, Cleburne, Columbia, Conway, Cross, Hempstead, Independence, Lafayette, Lee, Lincoln, Madison, Nevada, Newton, Ouachita, Phillips, Randolph, St. Francis,

During the 1940s and '50s, Game and Fish personnel captured and transplanted deer to help restore the state's whitetail herd. Few could have predicted the incredible increase in deer populations that occurred over the next fifty years. Today's Arkansas deer herd numbers almost one million animals, up from an estimated thirty-two thousand in 1946.

Arkansas Game and Fish Commission.

Stone, Union, Van Buren, and Woodruff Counties. Parts of Washington, Jefferson, and Hot Spring Counties also were opened. A move to prohibit the use of dogs for hunting deer in Independence County failed.

At its next meeting, the agency "took under consideration" a number of requests and petitions it had received asking that certain areas be closed again and that certain restrictions be placed on the use of firearms or guns in open areas, particularly those near more populated communities. It seems not everyone appreciated the annual fall ritual of booming guns as deer hunters poured into Arkansas's woods.

By the mid-1950s, wildlife columnist Apple, on November 13, 1955, was able to advise his readers that,

Above right, **Men about to dart deer**

Above left, **Darted deer being examined**

Right, **Deer being released**

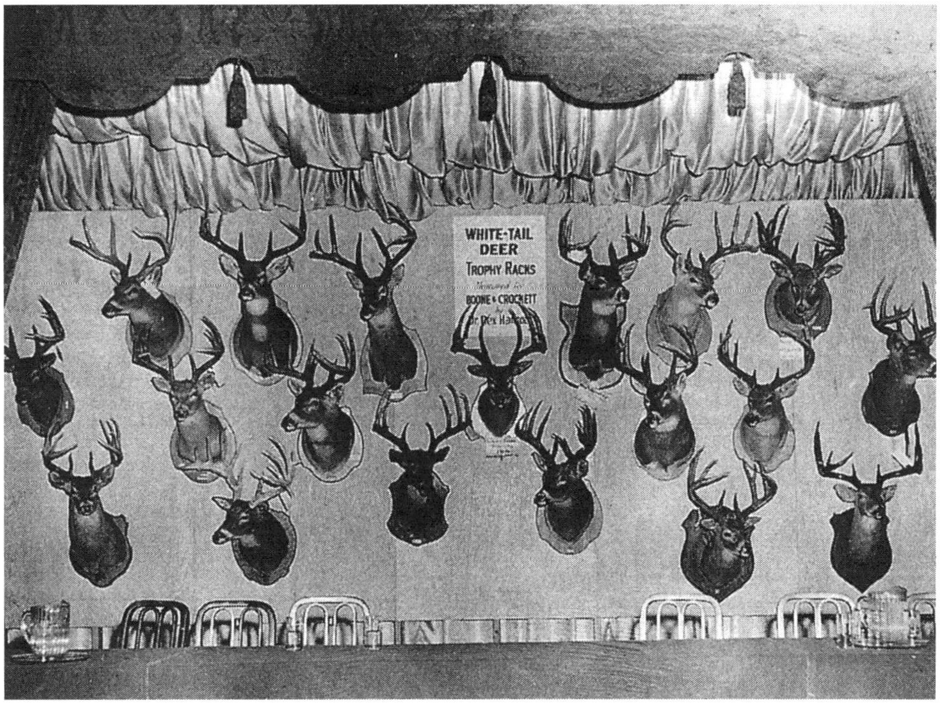

Display of trophy deer mounts

Proper management allowed Arkansas's deer population to recover to a level of abundance and quality in the 1950s. In 1962, Dr. Rex Hancock of Stuttgart, an official measurer for the Boone and Crockett Club, spearheaded a drive to find, measure, and enter record racks of deer antlers from Arkansas. Over one hundred racks qualified for the record book under the point system then in effect, including the nineteen pictured here. Most of these deer were killed during the 1950s.

Arkansas Game and Fish Commission.

"Prospects for this season's deer hunting in Arkansas appear to be among the best in our memory and should provide topnotch thrills for all sportsmen . . . According to preliminary reports, most of the deer will be killed in the southeast and south central part of the state. Izard, Stone, and Baxter counties have a good average kill each year along with some of the southwest portions of the state." By the mid-1970s, the deer population had increased to the point in southern Arkansas there was a disease outbreak thought to be associated with population density (Pledger 1977). An estimated 15 percent of the herd was lost to "epizootic hemorrhagic disease." The contagious malady is spread by biting midges characteristically in late summer or fall.

The increasing popularity of two forms of hunting—archery and muzzleloading—caused the commission to set special deer seasons for these weapons. Apple was a fan of bow hunting. "The sport holds great possibilities in the 31-day season" that opened October 1, he wrote on September 23, 1956. He continued,

> Arkansas has an abundance of deer and, in some spots set aside as special bow-hunting areas, any sex or size may be taken.
>
> Bow hunting season will be exciting, and it will test the hunting ability of the participants. Unlike hunting with a rifle or shotgun, every known trick of the Indian must be employed to bring the quarry within close range.
>
> Range for bow hunters is normally within 40 yards, as compared to 60 for the shotgun and much longer for the rifle. It is truly a sport where skill in stalking and marksmanship will mark the successful hunter.

Internationally renowned wildlife specialist Durward Allen commented in one of his books that the history of American game conservation has been marked by many a bitterly fought controversy, but "for heat and lasting quality the wrangle over our deer herds is near the top. From California to Maine and Texas to Ontario it has smoldered and flared . . . On a broad front the situation still is unsettled and

Archer with buck

Arkansas's first special deer season for bowhunters ran through the month of October in 1956.

Arkansas Game and Fish Commission.

the old issues regularly revive. Every state and the border provinces of Canada have deer 'situations.'" And so it was in Arkansas. An old issue that revives regularly concerns hunting deer with dogs. Apple definitely was not a fan of the practice. The commission, he noted, had run into some "pretty stiff opposition" with its regulation in 1954 to close the Ozark National Forest to deer hunting with dogs. Replying to the commission's critics, on September 5, 1954, he wrote,

> the basic reasons for the regulation are the long range benefits that will be attained through increasing our deer herd in this area. Other states that have banished the dog from the deer woods, as well as three areas in the Ozark region, substantiate these recommendations.
>
> The Sylamore and the Black Mountain areas both are definite proof of the high potential of increase that our deer herds possess when not continually harassed by dogs.
>
> It might be worthwhile for a lot of hunters to consider which is the greater sport—the adventure and craft necessary to hunt without dogs when only the hunter and his gun are matched against the deer and its environment or to have a pack of highly trained hounds run a deer out to the road for the man to shoot who is lucky enough to have the right stand.

Personnel from the University of Arkansas Zoology Department began a three-year research project on the effects of dog harassment in September 1971. The project compared experimental chases conducted in Drew County, which had a kill the previous year of 1,739, and in Madison County, where the kill was only 530. No deer were caught by the dogs during the experimental chases, a situation that often is not duplicated in real life. The preliminary finding, based on the experimental chases, was that there was "no evidence of detrimental changes [to the deer] in behavior or other ill effects."

The issue flared again in 1985 when the commission banned hunting deer with dogs during the December season in the mountainous areas. Opponents said the decision was arbitrary and not backed by sufficient evidence. By then, however, wildlife biologists had pinpointed what they considered to be the real problem, and it was not the dogs per se but the hunters with the dogs.

Steve Cole, Wildlife Management Division chief, explained, "There is a vast amount of research which supports . . . that using dogs for deer hunting in mountainous regions is too efficient a method of harvest. In fact, it is so widely recognized that this method of deer hunting is harmful to the deer herd that Arkansas is the only southeastern state left which still allows the use of dogs for deer hunting in the mountains" (Spencer 1984).

The use of dogs for hunting has been defended vigorously through the years by, among others, James E. Parrish, president of the Arkansas Fox and Wolf Hunters Association. He wrote in 1970 that hunting with dogs was one of Arkansas's "soul-satisfying" ways of life and that the state was not yet too populous to prohibit it continuing. Parrish equated the "soul-satisfying" way of life with dollars and asked his readers to multiply the 131,000 dog licenses sold by the commission in 1969 by the estimated $100 a year it took an owner to keep a hunting dog in good health. "You get $13 million generated in business each year by the hunting dog owners in Arkansas," he concluded.

One of the major challenges Arkansas wildlife biologists encountered in the 1970s and 1980s was undoing the education hunters had gotten and persuading them it was now time to harvest deer of both sexes in southern Arkansas to maintain a healthy herd. In the September 1986 issue of *Arkansas Game and Fish,* research coordinator Michael E. Cartwright said:

> The Gulf Coastal Plain region of south Arkansas contains the highest deer population in the state. As much as 60 percent of the statewide deer harvest has come from this region in recent years. Deer densities exceed 30 deer per square mile over much of the region and herd condition is poor.
>
> Research data on south Arkansas' deer herd indicates imbalances in herd sex ratios (too many does for the number of bucks), poor physical condition (low body weights, low quality antlers), and poor herd age structure (very few adult bucks over 2.5 years of age). These problems are a result of heavy pressure on bucks with predominately bucks only hunting seasons. As high as 70–80 percent of the legally harvested bucks are subadults at 1.5 years and some 2.5-year-old bucks have nubbin antlers less than 2 inches in length. This makes these animals illegal under legal buck definitions.

Either-sex deer seasons were instituted first in southern Arkansas, but this put added pressure on the commission to determine seasons. For example, commission chairman Tommy Sproles of Cato worried in 1990 that a short either-sex season was causing too many hunters to go into the woods at the same time, "presenting a dangerous situation." Game and Fish Commission officials like Wildlife Management Division chief Don Akers insisted successfully, however, that "Hunting both sexes of deer is the proper way to manage established deer populations. Harvesting does and bucks allows (1) the maintenance of a healthy, productive deer herd in balance with the available habitat, (2) a desirable sex ratio of one buck to two or three does, which is in line with natural population ratios and (3) the antlerless part of the herd to be utilized through legal hunting."

Legal Deer Harvests

YEAR	TOTAL	YEAR	TOTAL	YEAR	TOTAL	YEAR	TOTAL
1938	203	1954	7,343	1970	26,017	1986	79,888
1939	540	1955	6,856	1971	24,720	1987	106,392
1940	408	1956	8,249	1972	32,087	1988	110,207
1941	433	1957	9,438	1973	33,794	1989	113,079
1942	1,000	1958	9,993	1974	33,763	1990	90,910
1943	1,723	1959	12,290	1975	33,322	1991	110,896
1944	1,606	1960	15,000	1976	37,790	1992	110,401
1945	1,687	1961	19,359	1977	29,109	1993	122,063
1946	1,661	1962	27,772	1978	43,452	1994	120,491
1947	2,016	1963	25,148	1979	36,074	1995	163,924
1948	2,779	1964	16,637	1980	45,202	1996	152,460
1949	3,075	1965	17,138	1981	44,591	1997	not available
1950	4,122	1966	20,028	1982	42,873		
1951	4,600	1967	21,751	1983	60,248		
1952	6,090	1968	20,063	1984	66,039		
1953	6,245	1969	25,696	1985	60,076		

Source: Wildlife Management Division, Arkansas Game and Fish Commission, Little Rock, Ark., 1996.

This is how bobwhites, wild turkeys and white-tailed deer have fared in Arkansas since Amendment 35 was adopted. There are many other game species worthy of attention, however, even though some are not hunted. Before focusing on these, it is appropriate to explore efforts the Game and Fish Commission made through the years to improve its wildlife management.

For example, the independent study team that studied the Game and Fish Commission in 1970 made recommendations concerning the agency's wildlife management efforts. Among them were

> That the state upgrade its wildlife management and biological staff by increasing the number of trained professionals as rapidly as possible. To accomplish this, higher salaries will have to be paid. We recommend higher educational and professional requirements so that Arkansas will compare more favorably with other states;
>
> That cooperative programs of research, training, and conservation education be entered into with appropriate state colleges or universities;
>
> A much wider cooperative program with other state and federal agencies having responsibility of land use and management in Arkansas;
>
> That the commission give special attention to the Delta region of the state to the end that unnecessary drainage, clearing, and stream channelization be pre-

vented where possible and that steps be taken to minimize the ill effects on wildlife of such drainage, land clearing, and other habitat destruction as may be accomplished;

> That the commission give greater emphasis to its wildlife research program, with periodic review and evaluation of all research and management programs;
>
> Assignment of a staff forester in the Wildlife Management Division to manage all of the department's forest lands with fish and wildlife objectives; and
>
> That the state recognize its responsibility for nongame species.

Most of these recommendations have been carried out, at least to a limited extent.

In the late 1980s and early 1990s, the commission took three specific, significant wildlife actions: First, it adopted a regulation in March 1987 making it illegal to hunt within fifty yards of the center of any city, county, state, or federally maintained road during any firearm deer season. The exception was small game hunting using shot no larger than number 4. Enforcement Division chief David Herman urged the passsage of the regulation, saying there was "a mandate from the public." Second, the commission adopted a system under which the accumulation of 18 points for hunting and/or fishing violations in a three-year period would result in the revocation of all hunting and fishing licenses and privileges for a year. Accumulating 30 points in a five-year period would bring the loss of licenses and privileges for three years. (The point system was toughened in 1995.) And third, the commission adopted a replacement value (restitution) schedule for wildlife killed wantonly or by accidents such as toxic spills.

GAME WILDLIFE

MOURNING DOVE

For many Americans, summer ends and fall begins unofficially with the Labor Day holiday. Not so for Arkansas hunters. Though the calendar does not agree, fall begins for them with the opening of dove season, and like going to a deer or a duck camp, a trip afield often becomes a major social event for those eager to "burn powder and get their juices flowing" after a long, hot summer.

While usually plentiful in Arkansas, the dove also is a challenge to hunt, as Dick Gregory described in the September 1986 *Arkansas Game and Fish*:

> They're called Mourning Doves because they sit around all year making those cooing sounds that mimic mourners at a funeral. They flit from tree to tree, gather a little grit from your driveway, and raise their young in your backyard, never giving a clue to their secret lives as game birds.
>
> Come September, they gather into large concentrations for their migration and metamorphosis. Like Gremlins when wet, doves in September are a different creature. They evolve from mild mannered city-dwellers into twisting, diving, rocket-fueled transients that strafe hunters, and only occasionally succumb to the hunter's antiaircraft barrages.

Mourning dove wanted poster

In the 1960s, the Game and Fish Commission urged hunters to watch for leg bands on mourning doves they killed.

Arkansas Game and Fish Commission.

> They sneak in low over the horizon and skim the grain fields, low enough to swat with a fishing pole. They fly in squadrons of four or five, just out of gun range, getting the coordinates to drop a load of napalm. And, when you finally give up and return to your truck with your gun unloaded, a few of them will fly over in the "missing man" formation, dipping their wings and gloating over your lack of prowess with a scattergun.

Called the turtle dove by old-timers, the mourning dove is one of the most abundant and popular gamebirds in the nation and is hunted in more than thirty states. In states where it is not hunted, the dove is protected as a songbird. As Gregory indicates, the dove is one game bird that does not require much habitat management because of its adaptability. Most land use practices in Arkansas attract and hold mourning doves because, though they seem to prefer grain food, they will eat corn, peas, soybeans, and enormous quantities of weed seeds. Milo was a preferred food, but new bird-resistant strains have cut back on its use as a food. Doves often are seen sitting on roadsides and gravel and sand bars where they take dust baths and pick up small bits of rock and grit that their gizzards need to help grind the seeds.

A 1984–85 Game and Fish Commission survey showed 450,000 to 670,000 doves were harvested in Arkansas during that season. About thirty-five thousand Arkansas hunters participate in the sport, putting it on a par with quail hunting and just behind waterfowl hunting in popularity.

Because of their abundance and the assumption that they will always be plentiful, doves seldom get mentioned at Game and Fish Commission meetings or hunter public hearings. An exception occurred in August 1994 when commission biologists reported that there was a 20 percent increase in the dove population, which they attributed to plentiful spring and summer rains that brought about above-normal seed production.

RUFFED GROUSE

In September 1948, the Game and Fish Commission voted to amend its federal-aid project plan to buy one hundred ruffed grouse from the Sandhill Game Farm in Babcock, Wisconsin. The grouse were to be planted in the Black Mountain Refuge in Franklin County and the Sylamore District in Stone and Baxter Counties. A resolution to "completely close" the season for hunting grouse for an indefinite period followed. In October 1949, the commission renewed the restocking program to acquire one hundred or more grouse, and it took similar action in 1950.

Though they knew grouse are highly susceptible to several different diseases, biologists did not know why grouse disappeared from the Ozarks in the late 1800s. Destruction of habitat, fires, vegetation changes, and intensive grazing were suspected causes. In addition, old settlers recalled a prolonged ice storm that might have delivered the *coup de grace*. At any rate, conditions—especially the general cessation in wildfires—had stabilized enough by the 1950s, particularly in the Ozark National Forest, to warrant the expense and time of a restocking effort. About three hundred grouse were released (Holder 1951). Some survived, but the stocking did not take over the long-term.

The commission maintained its interest in restocking grouse, however, and breaks started to come in 1981 when resource managers at Buffalo National River learned they would be allowed to trap grouse at Cumberland Gap National Historic Park on the Kentucky-Virginia border. Missouri had started earlier on a grouse restocking program and was having considerable success. Experts from Missouri toured potential release sites in the Ozarks in 1981 and advised there was some excellent habitat in Arkansas—even better than some sites where grouse were released in Missouri (Pharris 1983).

A crew from the commission and the National Park Service went to Cumberland Gap and returned with fifty grouse that were released on Buffalo National River park lands in Newton County. Surveys indicated a good survival and reproduction rate. Meanwhile, a ruffed grouse restoration plan was prepared by biologists and approved by the commission in April 1982. It called for three hundred grouse to be released annually for five years. Though Arkansas biologists trapped grouse in Virginia and Minnesota for transfer to the state, finding sources of wild birds remained a major obstacle (Pharris 1983). Arkansas persisted, however, and through 1987, more than six hundred grouse trapped in Wisconsin, Minnesota, Kentucky, Ohio, New Jersey, Pennsylvania, and Virginia were released on five Ozark

A ruffed grouse from Wisconsin is readied for release in Franklin County's Black Mountain Game Refuge, 1949. About three hundred grouse were released in 1948, '49, and '50, but these stockings were unsuccessful. A new and more successful restoration effort began in 1981. Grouse were trapped in seven states and flown to Arkansas for release. The progeny of these birds now inhabit several areas in the Ozarks.

Arkansas Game and Fish Commission.

Above, **Game and Fish employee Richard Bearden picking up grouse in boxes at airport, 1988**

Left, **Ruffed grouse ready for release, 1949**

sites. Released birds were monitored year-round by the commission and cooperating agencies. Summing up the success of the restocking, Game and Fish biologists Ron Smith and Mike Cartwright wrote the following in September 1987 for the agency magazine:

> Today, a sound absent in Arkansas for more than 60 years can be heard once again. A hollow thumping comes from a hidden spot in the woods where a male ruffed grouse begins his mating ritual in late March. Rapid wing movements create the drumming that sounds like someone trying to crank a diesel tractor on a cold frosty morning. The male's drumming advertises its territory to other males and attracts females for mating.

The number of grouse restocked in Arkansas rose to twelve hundred by 1994 at a cost of about two hundred dollars a bird, and at least that many more were said to be needed. Success varied. Biologists say grouse may succeed but never be abundant because Arkansas is the southern tip of the bird's range. Missouri took twenty years to have a grouse season; Arkansas could have a limited hunt by the year 2000.

CROWS

Crow hunters in Faulkner County, 1995

Modern hunters use a variety of specialized calls and decoys to attract their quarry.

Keith Sutton.

American crows occur throughout most of the continental United States, living in a variety of habitats, including farmlands, woodlands, and even urban areas. They seem particularly fond of large open farmed fields where they can do serious crop damage, which explains why they are not welcomed by farmers.

Observations from the early 1900s suggest crows were not abundant in Arkansas then, but their numbers increased as lands were cleared for intensive farming. Crows are actually omnivorous, eating everything from spiders, snails, snakes, eggs, and worms to grasshoppers and other harmful insects. They also are important scavengers, helping to clean up road kills, dead fish, and other carrion, and they hunt assiduously for wild bird eggs and nestlings, which makes them a favorite target for hunters who prefer other quarry or just want to protect songbirds and herons.

The Audubon Society's *Encyclopedia of North American Birds* tells more about the crow, noting that experiments with captive American crows "showed that they can count up to three or four, are keen, wary birds, good at solving puzzles and performing astonishing feats of memory, and quickly learn to associate various noises and symbols with food . . . and it has been known to naturalists for years that they have a language of their own." One study describes twenty-three calls and their meanings.

Fish crows favor coastal marshes, beaches, swamps, rivers, lakes, and other wetland habitats where they forage for crayfish, turtle eggs, fruits, and dead fish. In Arkansas, fish crows usually are seen near major rivers, especially in the lowlands and in the Arkansas River Valley where they live year-round. Though common in some localized areas, sometimes forming flocks of fifty or more birds, fish crows are much less common in Arkansas than American crows.

Hunting is a means of controlling crow populations. There is no bag limit in Arkansas during the season that runs through most of fall and winter.

DUCKS

The Arkansas Game and Fish Commission does not get as much flak as it ordinarily would from the state's ardent duck hunters because hunting regulations, including season parameters and bag limits, are determined by the U.S. Fish and Wildlife Service under the international migratory waterfowl treaty discussed in part 2.

Waterfowl seasons and bag limits are not set arbitrarily or on the basis of some hunch by federal bureaucrats. Extensive low-level aerial surveys over breeding grounds in the northern United States and in Canada, post–hunting season surveys, and a banding program are the basis for the service's annual regulations. When the agency feels it has a handle on how the birds have fared on their breeding grounds, the information goes to the four Flyway Councils, which are made up of the directors and designated biologists from each affected state's wildlife agency. Arkansas is in the Mississippi Flyway Council.

Tenderfoot hunters or nonhunters may think they are seeing a lot of ducks in east-central Arkansas in late November and December, but the flocks are but a fraction of those described in part 2. A strong vein of pessimism ran through virtually everything that was said in interviews or written about waterfowl for almost half a century. The pessimism was based on the seemingly irreversible loss of habitat; optimism did not begin to return until the mid-1990s as a result of several initiatives, including a concerted three-nation effort to preserve and restore duck habitat.

"The fight to keep a supply of ducks coming south to Arkansas is a never-ending one," Bill Apple observed on October, 27, 1957. Persons like the late Dr. Rex Hancock of Stuttgart, who fought for more than a decade to keep the Army Corps of Engineers from ditching the Cache River's and Bayou DeView's prime duck habitat, undoubtedly agreed.

On April 10, 1955, Apple confessed that the future of duck hunting in Arkansas "has never looked so poorly to me as it does now. If there is any one thing clouding up the duck horizon it is the intensive drainage of our wetlands which is siphoning off our future duck production just as surely as an open wound would drain away the life blood of a person." Apple said the duck population was about 200 million in the 1920s but had dropped to 30 million in the 1950s.

He estimated that a 217,000 square mile area of pothole (glacial ponds) country in the United States and Canada furnished breeding habitat for about one-third of the spring duck population of North America. Twenty-five percent of the area is below the Canadian border, primarily in the Dakotas and Minnesota. "Continued high production from this region was regarded as essential to the future waterfowl hunting throughout the Mississippi." Man invaded the breeding grounds of the migratory waterfowl in Canada during the 1920s, draining vast areas of water. The

W. H. Claypool with yellow Lab

Claypool, a world champion duck caller, developed a duck sanctuary and hunting club near Greenfield, Arkansas. In 1956, millions of television viewers across the nation saw his Wild Acres duck preserve when it was featured on NBC's "Wide World." An outstanding conservationist, Claypool died in 1973.

Arkansas Game and Fish Commision.

effects of this action prompted the formation in 1937 of Ducks Unlimited. Duck stamp sale funds collected in the United States could not be used outside of the country, so sportsman had to raise the money to salvage what they could of the Canadian breeding grounds. Ducks Unlimited has been in the forefront of this effort.

Pothole drainage goes on in the northern United States, just as draining and clearing of wetlands continues in Arkansas, except when there is a severe economic downturn such as farmers experienced during the mid-1980s. Loss of breeding habitat was not the only explanation for Arkansas's reduced waterfowl flocks. Arkansas faced stiff competition from other states in the Mississippi Flyway. As rice growing

Claypool ducks

Patricia Peacock, world champion duck caller

Pat Peacock Stephens (now Pat Stephens Johnston), appointed to the Arkansas Game and Fish Commission by Gov. Jim Guy Tucker in 1994, was the first woman to serve on the commission. She is seen here hunting ducks near her home town of Stuttgart in 1955.

Arkansas History Commission.

began moving farther north during the 1950s and 1960s, more man-made hunting areas developed. As more attractive water was created northward along the Mississippi River in Illinois, Missouri, and Iowa, the duck population became decentralized. The Tennessee River with its impoundments also diverted many of the Mississippi Flyway ducks.

Carl Hunter, former assistant director of the Game and Fish Commission, was pessimistic in 1959 when he was wildlife manager for Wingmead Farms at Roe in Monroe County. Hunter wrote in the *Arkansas Democrat* that those who longed to see the vast flights of ducks over Arkansas return were in a "losing battle." He was not suggesting ducks would go the way of the buffalo, he explained, "but a large decline is certain and the day of free public hunting is fast coming to a close." In an article published in the *Arkansas Democrat* (January 4, 1959), Hunter went on to explain why he held this view:

> Since the days of shocked rice and winter long harvesting (only about 15 years ago) newer combines have cut losses of rice in the field to less than 5 percent even under adverse harvesting conditions. The end of September finds only a rare field of uncut grain. Modern machinery moves in, turning under the stubble, and with it much of this small percentage of shattered grain.
>
> Our vast natural feeding ground, the bottom-land timber, is flooded only irregularly. Straightening and channeling of the rivers, construction of the large dams, ditches and levees has reduced the needed overflow and made it more erratic which discourages feeding in these flat lands. Much of this area is being cleared.

Either Hunter had long-range vision or the conditions he described got worse instead of better. Mrs. Bennie Burkett, a lifelong Grand Prairie resident, and Dee Simpson, who began his commercial hunting and fishing guide career in 1926, expressed almost identical views during interviews in December 1985. Modern harvest equipment is so efficient that little grain is left in the field on which the ducks can feed, they said. Further, the equipment also is speedy, so that by the time the ducks arrive, what grain has been left already has sprouted because the harvest was completed more than two months before.

As described in more detail in part 3, an international agreement by Canada, the United States, and Mexico called the North American Waterfowl Management Plan went into effect in 1987 and was augmented in 1989 by congressional passage of the North American Wetlands Conservation Act. The plan's goal of bringing back duck and geese populations—6 million geese and 62 million breeding ducks supporting a fall flight of at least 100 million birds under average environmental conditions—to at least 1970 levels seen seemed ambitious at the time but has encountered early success with help from the replacement of lead shot with steel shot and with the federal Conservation Reserve Program.

Arkansas duck hunters began seeing results from these efforts and those of Ducks Unlimited in December 1993 when Game and Fish Commission waterfowl biologist Tim Moser announced aerial surveys showed "More mallards, more total ducks, more snow geese and more white-fronted geese . . . than in any other waterfowl survey since 1985, when the current survey system was adopted." The survey found 1,061,869 ducks of all types in Arkansas, including 751,695 mallards, the species preferred most by hunters.

The news kept getting better. The Mississippi Flyway Council, of which Arkansas is a member, was limited to a thirty-day, three-duck season until 1994. Better breeding-ground conditions, aided by increased rainfall, brought more ducks in 1994, so the season was lengthened to forty days, but still with a three-duck limit in Arkansas.

In mid-1995, the U.S. Fish and Wildlife Service reported breeding ground surveys showed duck numbers were up again—way up—for the Mississippi Flyway. Mallards had climbed to 8.3 million, up 8 percent over 1994 and the highest level

Arkansas's first duck stamp

Renowned wildlife artist Lee LeBlanc of Iron River, Michigan, was commissioned to paint Arkansas's first state duck stamp, which was issued for the 1981–82 waterfowl season. The scene shows mallards lighting in a wooded area on Bayou Meto.

Arkansas Game and Fish Commission.

since 1972. Canvasback ducks climbed 47 percent to 771,000. Redheads were up 36 percent to 888,000, and gadwalls were up 22 percent to 2.8 million. Overall, the breeding ground population of ducks was 35.9 million, up more than 10 percent from the year before. It was enough to give Arkansas hunters a fifty-day, five-duck–daily-limit season.

Another fifty-and-five season was in the works for Arkansas duck hunters in 1996 because midsummer reports from the U.S. Fish and Wildlife Service put fall flight forecasts up to 89 million, the highest since 1971. Mallards were up 10 percent over the long-term average, and mallard breeding populations were 7.9 million, close to the North American Waterfowl Management Plan's goal of 8.1 million. Major increases were noted as well in the numbers of blue-winged teal, green-winged teal, and northern shovelers. Canvasbacks, troubled for years by declining populations, were 56 percent ahead of the long-term (since 1955) average, and redheads were up 42 percent. American wigeons, scaups, and northern pintails showed decreases, however.

Mallard numbers in southern Saskatchewan, where most mallards that winter in Arkansas breed, were up by 18 percent in 1996 over the year before. The federal agency estimated 584,200 mallards were taken in Arkansas in 1995, nearly one-quarter more than the previous year and third only behind 1975 and 1980 in modern duck seasons.

Arkansas's total duck take in 1995 was just over one million, the U.S. Fish and Wildlife Service said, an all-time high since such statistics first were gathered in 1961. Of the total, 58 percent were mallards, 16 percent gadwalls, 8 percent green-winged teal, 6 percent wood ducks, and the remainder other species.

CANADA GEESE

Commission biologist Dave Donaldson observed in 1968 that the great flocks of majestic Canada geese that once wintered along the White and Arkansas Rivers had dwindled to "near nothing." Based on aerial midwinter inventories, the Canada goose population on the White River National Refuge had declined gradually since 1947 from fifteen thousand to about two thousand in 1968. In 1950, the Canada goose population in Arkansas took a decided drop. From 1951 to 1954, the flock remained fairly stable and then dropped again in 1955.

Donaldson speculated that the population in the 1950s was overharvested. Kill estimates during those years indicate eight hundred to twelve hundred birds, most of them probably young geese, were taken annually by Arkansas hunters, and this represented the annual reproduction effort. Another possibility, Donaldson wrote, was that habitat on and around the White River refuge was supporting all the geese it could. Large-scale habitat improvements for geese did not begin until 1955 and were not in full swing until 1957. Since then, the population on the refuge has been so low that geese have not been using all the grain and browse provided for them.

Donaldson also cited the possible "shortstopping" effect of federal and state refuges in states north of Arkansas, primarily Missouri, Illinois, and Wisconsin. Illinois, starting with the acquisition of Horseshoe Lake near Cairo in 1927, had developed a series of refuges primarily for Canada geese. Illinois also indulged in what others considered dirty tactics at the Horseshoe Lake refuge during the late 1950s. The day after the hunting season closed, the refuge lands and others around it were sprinkled with tons and tons of feed. "No goose in its right mind would leave all that behind," outdoor writer Henry Reynolds commented. U.S. Fish and Wildlife Service officials denied at a meeting in Memphis that any baiting was conducted at the refuge during hunting season, which prompted Reynolds to quip, "If all the corn left on the ground in and around the Horseshoe Lake Refuge got there through natural agricultural procedures, then John Deere and International Harvester had better revamp their corn pickers." At any rate, Donaldson said there was no doubt in his mind that the presence of large wintering flocks on the new refuges to the north had some effect on Arkansas's Canada goose population.

In 1962, the Fish and Wildlife Service said if Arkansas and Louisiana would agree to a closed season on Canada geese, it would transport live trapped geese into federal refuges in the two states. Both agreed, and in 1962, some 630 geese trapped at Swan Lake in Missouri were released on White River refuge. The transplant program was repeated in 1963, 1964 and 1965 with 1,500 birds from Swan Lake being moved into Holla Bend National Wildlife Refuge near Russellville and fewer geese from Horseshoe Lake going into Wapanocca National Wildlife Refuge near West Memphis. Five years after the transplant program started, with some five thousand birds having been moved to White River and with a closed season, the wintering population was nearly one thousand birds, fewer than it was when the program began. Somewhat better results were seen at Holla Bend, but even there, the program was not considered successful.

Donaldson saw what he called a "glimmer of hope" in a program that Carl Hunter had carried out at Peckerwood Lake on Wingmead Farms. Hunter obtained live-trapped birds that had been hatched the previous spring, were on the wintering grounds for the first time, and had not been imprinted to any wintering area. He clipped the primary flight feathers of the birds, as had the commission biologists on the White River and Holla Bend transplants, but Hunter kept them clipped for three years, until the birds had reached sexual maturity. Then he allowed the birds to become free fliers. Most of the geese migrated north, and many returned, bringing their broods with them. He eventually built a flock of two thousand geese that annually used Wingmead.

The season on Canadas was closed in 1979 to protect another restocking effort the commission had in mind and on which it began working in earnest in May 1981. The latest program was based at the Cummins Prison Farm because of its proximity to highly developed farm lands where waste grain would provide adequate food for migrant geese. The program was similar to the one tried earlier in that it involved trapping geese in Illinois and Missouri. This time the birds went into holding enclosures near Grady in Jefferson County. The latest program was distinctly different in three ways from the previous one: First, only yearling birds were brought to Arkansas, so they would not have been imprinted to another wintering area. Second, each site was functional for three years, guaranteeing exposure of the geese to the areas. Finallly, only the early migrants were transplanted because banding studies indicated they were related to others using the more southerly area of the Mississippi Flyway (Walter 1982).

Giant Canada geese in a commission breeding facility near Clarksville

Birds hatched at this location were released to help establish a resident flock of giant Canada geese in western Arkansas.

Gregg Patterson.

The last of 403 geese was released in 1985. Success of the project was to be based on a 50 percent return of marked birds to the area. In 1988, an aerial count of Arkansas's resident giant Canada geese showed 175 breeding pairs, about three times the number counted just three years earlier. Hunting the popular and colorful waterfowl was resumed in eastern Arkansas on a limited basis.

The commission also tried to build the population of interior Canada geese in western Arkansas by releasing a captive flock of migrant geese near Russellville into the wild. The hope was that the birds would return to the area each year, bringing their newly hatched young with them.

Tim Moser, the Game and Fish Commission waterfowl biologist, estimated conservatively in 1992 that from three thousand to forty-five hundred Canada geese lived year-round in northwest Arkansas, some on small farm ponds and others along the Arkansas River and on the big lakes. When other geese migrated from the north, as many as twenty-three thousand Canadas were present in the area, he said. On his recommendation, the Game and Fish Commission reopened Canada goose hunting on a limited basis in 1992.

It is difficult to estimate how large Arkansas's population of migratory Canada geese is now because residents mix with varying numbers of the birds from midwestern states in any given year. Counts taken in early December, before the weather gets cold enough to force large numbers of geese to retreat from northern states, indicate Arkansas probably has some ten thousand resident migratory Canada geese.

SNOW GEESE

The snow goose is a highly sought game bird over most of its migration and wintering ranges, but the relatively recent occurrence of large populations wintering in Arkansas has not generated a high demand by state waterfowl hunters. Most of the

Snow geese, Arkansas County, December 1994

Theresa Sutton.

three hundred thousand snow geese harvested nationally each year are killed in Texas and Louisiana. Birders recognize the aesthetic value of snow geese, however, and they trek to Holla Bend and White River national wildlife refuges to see and photograph the great gatherings of snows.

The birds usually begin arriving in Arkansas in early November, with the major concentration coming in January. They begin returning to their northern breeding grounds during the first warm days of February. The abundance of irrigation reservoirs and rice fields flooded for duck hunting have proven ideal for snow geese.

Joe Mosby, outdoor editor of the *Arkansas Gazette* until its demise in 1991, said a daily bag limit of two mallards was just not enough for many duck hunters, and he urged them to try goose hunting:

> You hear before you see. A nasal "squonk, squonk" is coming your way, and the flock of incoming snow geese stretches wide over the horizon. You wait, you tense muscles, you test your patience . . . Your face is down on the ground, and finally the guide or leader of the party yells "take 'em" and you roll over and come up with the shotgun.
>
> It is another memorable experience. If you haven't had this pleasure, it can be a goal, a quest for you."

Snow goose populations began mushrooming in the 1990s as the growing of winter wheat expanded in Arkansas from a casual crop between plantings of soybeans to a major endeavor, particularly for northeast Arkansas farmers. The influx of snow geese came shortly after the increase in wheat farming. Biologists began to worry that snow geese, with few natural predators about, were becoming so numerous that they were causing their breeding-ground habitat in the far north to deteriorate. Certainly the snows were beginning to worry Arkansas wheat farmers. In response, the Game and Fish Commission extended the hunting season on snow geese and urged farmers to open their lands to goose hunters.

GREATER WHITE-FRONTED GEESE

A little larger than a snow goose but a little smaller than most Canadas, white-fronted geese, also called specklebellies, migrate into Arkansas earlier that most of their cousins. They also move farther south earlier when really cold weather sets in. Historically they used the Central Flyway to California, but waterfowl biologist Moser said a shift began to the Mississippi Flyway in the 1990s, with white-fronts surpassing Canadas in wintering populations in Arkansas.

Since 1990, an annual average of 40,000 white-fronted geese, compared to 2,000 to 3,000 before, spent December in Arkansas before moving farther south in January. Canadas number only about 13,000 in December. From 1971 to 1975, Arkansas hunters bagged an average of 200 white-fronted geese annually. From 1981 to 1985, the average picked up to 1,000 a year and then soared to 5,000 birds annually in the early 1990s.

White-fronts make more use of soybeans, including cutover fields, than other geese, although rice is their preferred grain among agricultural crops.

White-fronted geese, Canada geese, and snow geese gather in separate groups in the far north and then travel to Arkansas separately. Once in the state, they often use the same feeding areas but tend to stay in their own clans, with Canadas keeping a wider distance from the snows than do the white-fronts. Geese also have their own languages or honks. White-fronts are said to yodel with a distinctive "*wa-wa-wa-wa*" that causes some old-timers to refer to them as laughing geese.

White-fronted geese in flight

The number of white-fronted geese have steadily increased in Arkansas since the 1970s.

Keith Sutton.

Coot

Though considered a game bird, the American coot, which is common throughout Arkansas, is scorned by most hunters.

Keith Sutton.

AMERICAN COOT

The coot is the klutz of the bird world, plentiful yet ignored and even ridiculed by Arkansas hunters. They do not fly like the admired mallard, teal, or wood duck; a coot runs clumsily along the top of the water for yards before getting into flight on wings too small for its body. Yet the coot is prolific and did not suffer the same decline as ducks. When the government dropped the limit as low as two drake mallards a day, it allowed hunters to bag fifteen coots a day.

The coot is not a duck; it is in the rail family. Uniformly slate-gray or nearly black, the coot feeds on small fish, snails, and insects. Coots are common on most Arkansas lakes and rivers. A few live year-round in Arkansas. Many more migrate from the north when the weather turns cold, and populations are healthy.

BLACK BEAR

The Bear State was almost devoid of its unofficial namesake by 1900 and was down to between forty and fifty black bears in the 1940s, most of them a remnant of the original population in the White River National Wildlife Refuge. A halting step to reverse this was taken in 1949 when the Game and Fish Commission released four bears in the Ozarks—two in the Black Mountain area and two in Stone County (Wilson and Gipson 1975).

Ten years later, the commission decided to try in earnest to restock black bears by entering into an agreement with the Minnesota Game and Fish Commission and the Canadian province of Manitoba. Under this arrangement, Arkansas traded bass and wild turkeys for the privilege of trapping black bears for transfer to the state (Pharris 1981). From 1959 until the program ended in 1968, 268 bears were trapped and released in remote areas, primarily in the Ouachita and Ozark Mountains where habitat was considered excellent.

Bear traps were made of two fifty-five-gallon steel barrels welded together. Three

Bear release

Black bears from Canada were released in Arkansas in October 1964.

Arkansas Game and Fish Commission.

of the four ends were cut out. There was a "rabbit-gum" door at one end, and at the other, several holes were bored to provide ventilation. Fish, beef, and other meat scraps, plus various fruits and honey, were used as bait, but the best was bacon grease and honey (Rush 1967). Trapping was done at least five miles from the nearest human habitation. The program went so well that Arkansas's bear population was estimated to be between six and seven hundred in 1973 and between twelve and fifteen hundred in 1980. In 1994, the population estimate was twenty-two hundred.

Bear restocking ended in Arkansas when most of the better bear habitat became occupied. Arkansas's first bear season in modern history was held in 1980 after a fifty-three-year hiatus. A carefully planned, experimental, limited hunt, it was designed primarily to reduce the bear population to the ideal carrying capacity of the habitat (Rogers 1980). The hunt was restricted to four wildlife management areas: Piney Creeks between Jasper and Russellville; White Rock, which covers a large part of four northwest Arkansas counties; Buffalo River near Jasper; and Gulf Mountain near Clinton. Permits for the first two areas were limited to 2,000, while only 150 were issued for Buffalo River and Gulf Mountain (Wilson and Gipson 1975).

The hunt was set for December 1–6 to protect female bears, which usually den earlier than males. Hunters were allowed to take only adult bears—those weighing seventy-five pounds or more, excluding sows with cubs at their sides. Preseason predictions of the harvest never exceeded twenty. The 1,334 hunters who actually participated bagged five bears, not considered a good success ratio. The participants also discovered that bear hunting was hard work and that the skills last taught fifty-three years ago needed to be honed (Pharris 1983).

With the bear population stabilized between 1,500 to 2,000, special hunts continued. The 1985 harvest totaled 23 bears: the harvest rose to 102 by 1991 and 130 by 1995. Eleven nuisance bears were moved in 1985, primarily in response to complaints from beekeepers. Regardless of the effort to shift bears, it appeared a backlash might be building by the mid-1980s. For example, it was reported in the December 21, 1985,

Bear hunter and archery gear

Bud Flowers of Center Ridge took the first bow-killed bear in modern Arkansas history during the second day of the week-long 1982 bear season. The bear weighed 425 pounds and measured eight feet, four inches from nose to hind foot.

Arkansas Game and Fish Commission.

Arkansas Gazette that an Oklahoman was mauled by a bear in the Ouachita Mountains while raccoon hunting. The attack took place near Mena when the hunter's dogs apparently trailed the bear to its den. The bear was defending its territory when it attacked the hunter who entered the den. An Oklahoma wildlife officer said he was hearing more complaints about bears in the area, mostly about them attacking livestock. He said Oklahoma had done no restocking, so the bears probably came from Arkansas. "A bear needs a pretty large territory," the officer said. "People are moving into areas where they are." Even so, he added, he did not regard the bears as being any "real problem," and in fact, "We are trying to preserve some of them if we can."

Wildlife Management Division chief Steve Cole told the commission in 1986 that the practice of baiting bears out of season for viewing or scouting purposes was causing nuisance problems. The commission responded by making it unlawful to bait bears at any time, with a penalty for violation ranging from $25 to $250. Also to help defuse the situation, the commission told the public in 1987 what to do if a bear was encountered: do not startle it; slowly move away in the opposite direction, preferably upwind. In the May 1987 issue of *Arkansas Game and Fish,* staff member Dr. Joe Clark, one of the nation's leading bear biologists, assured the public that:

> With 2,000 bears in Arkansas, and hundreds of thousands of hunters, hikers and other recreationists in the field each year, the number of opportunities for contact between bears and people is enormous. Yet only a handful of conflicts are reported, and we haven't had anyone hurt by a bear in recent history. Bears and people get along remarkably well, really.

Bears actually have far more to fear from humans than the other way around: unnecessary and wanton killing of the animals proved disturbing during the 1980s. The most flagrant case occurred on opening day of the 1987 deer season when twenty-nine-year-old Richard M. Perin of North Little Rock killed two bears minutes apart on the White River Refuge and left their carcasses. These were two of the last native Arkansas bears. One weighed 450 pounds, the other 400. The state's law enforcement community was irate. Several rewards were offered, and U.S. Attorney Chuck Banks visited the Fish and Wildlife Service Office in Little Rock to say he would personally prosecute the case if the culprit was found.

Months later, an informant identified Perin as the bear killer to Game and Fish Commission district enforcement supervisor Jack Byrd. Members of Perin's hunting party were found and interviewed but would not talk. They subsequently were subpoenaed to appear before a grand jury, and the penalties for perjury were explained. Within days, Perin turned himself in and eventually pleaded guilty. U.S. Magistrate John Forster fined Perin the maximum one thousand dollars and sentenced him to one year in jail, suspended because he was a first-time offender. Perin was not allowed to hunt or possess a firearm during three years of supervised probation: he had to perform four hundred hours of community service work and report to Judge Forster what he had learned by reading a doctoral dissertation on black bears.

In 1996, the Game and Fish Commission began seeking public comment on whether there should be a limited, tightly restricted hunt in the lower White River area. Bears there had begun widening their territory to wooded lowlands in much of southeast Arkansas, and some farmers were complaining that they were destructive nuisances.

The White River bears are believed to be closely related to the Louisiana subspecies of black bears that have been given "threatened" status under the Endangered Species Act. Research on the subspecies has been undertaken in both Louisiana and Mississippi. Meanwhile, the restocking of black bears in west Arkansas remains the most successful reintroduction of any large carnivore in the world.

Legal Bear Harvests

YEAR	TOTAL	YEAR	TOTAL
1980	5	1989	30
1981	20	1990	19
1982	19	1991	102
1983	31	1992	44
1984	31	1993	115
1985	23	1994	126
1986	13	1995	130
1987	7	1996	207
1988	14	1997	not available

Source: Wildlife Management Division, Arkansas Game and Fish Commission, Little Rock, Ark., 1996.

ELK

As noted earlier, Arkansas once had elk from the Mississippi River to the Ozark and Ouachita Mountains, but settlement wiped out their habitat. There were at least two early attempts to bring elk back to Arkansas, both commercial ranching

Elk calf

In 1948, the Arkansas elk population was estimated to be twenty-five. But by 1954, only one remained, this calf near Huntsville in Madison County. It probably was a descendant of the final herd of eleven elk released in Arkansas by the U.S. Forest Service in 1933.

Arkansas Game and Fish Commission.

ventures. A community near Eureka Springs is known as Elk Ranch, once a stop on the bygone Missouri and North Arkansas Railroad where investors had hoped to ship elk to midwestern and eastern markets.

In 1981, the Game and Fish Commission began trading largemouth bass to Colorado for elk, and by 1985, Newton County had about ninety-five of the second largest member of the deer family. Nebraska gave Arkansas a bonus of seven elk from its surplus in November 1984. Robert Harrison of Jasper, a volunteer who helped move the Colorado elk, said they were "mean, wild and stout" but that he was "tickled to death" to have them in Arkansas. A Newton County wildlife officer said he was confident that the elk would make it, but the adaptation process would be slow. Steve Cole, the commission's Wildlife Management Division chief, said at the time that he doubted the herd would reach huntable size, but only because of a lack of "suitable habitat."

The Rocky Mountain Elk Foundation provided two-thirds of the $17,100 that was needed to help the commission create fourteen wildlife openings (cleared areas in the dense understory of a forest that encourage growth of new green forage) in the early 1990s as an elk habitat enhancement project on the Gene Rush/Buffalo River Wildlife Management Area near Hasty. Foundation and commission officials found evidence in 1992 that the elk definitely were making use of the sites. Like cattle, elk are grazers and not browsers like deer.

Northwest Arkansas residents, particularly those in Newton County, were comfortable with the elk, viewing them as "their own" and getting riled whenever one was wantonly slaughtered. Elk are strictly protected by law. One young bull elk became somewhat of a local celebrity, and Newton County residents raised more than four thousand dollars as a reward in December 1983 for information leading to the conviction of the person who killed it. Local residents were riled again in 1993 when the body of a mature bull elk weighing eight to nine hundred pounds was found in the middle of a gravel road west of Jasper. It had been shot twice with a high-powered rifle.

The elk herd, named in honor of its leading advocate, Commissioner Hilary Jones, was reported to number about 175 in 1989. Cole continued to be cautious, telling the commission there were potential problems "with the growth of the herd in relation to their size" and in transporting them for treatment of diseases. Herd size in 1992 was roughly two hundred and may have increased by about one hundred by 1994. In early 1996, the commission solicited public comment on a possible limited, tightly restricted elk hunt in the Buffalo River area. Director Wilson emphasized that it was "just an idea," and added, "There is no way we will open hunting [on elk] if it would be a threat to the overall population."

Bull elk near Ponca on Buffalo National River lands in 1995

Many people visit this area each year to see and photograph the elk herd.

Theresa Sutton.

RABBITS

Rabbits made it into the ranks of regulated wildlife in Arkansas on January 21, 1952, when the Game and Fish Commission, with a five-to-two vote, made it a violation to hunt rabbits "except during the regular fall hunting seasons on other game species." It fixed the dates for the rabbit season from September 15 to January 31 for that year. A regular resident hunting license was required, with a daily bag limit of eight and a maximum of sixteen in possession. Further, the commission made it a violation to export, ship, transport, or carry live or dead rabbits out of the state except on a nonresident hunting license with a shipping limit of eight per license.

On September 9, 1956, *Arkansas Democrat* wildlife columnist Apple observed that Arkansans had put the rabbit on the bottom of the list of choice game. But in many states, particularly the northeastern section of the country, it is the favorite, and more ammunition is burned on the rabbit than any other single species. By his own admission, Apple did not have much regard for rabbits until he discovered how much fun it was to hunt them with beagles. When the beagle made its appearance, Apple said,

> it brought us a new sport that can only be appreciated by an ardent fox hunter.
>
> While the deep voice of the fox hound is missing, the cheerful yipe-yipe of the beagle is equally thrilling. Its value can hardly be underestimated, and its small size makes it ideal for the sport.
>
> Killing the rabbit is only a small part of the sport, and actually more enjoyment is derived from listening to and watching the beagles work. It's a lazy way of hunting, for a rabbit has a habit of circling, and one has only to rest and wait until the hounds have brought him back to the starting point.

The Game and Fish Commission said in its magazine in 1967 that rabbit hunting "is one of the state's important sports," but that while many hunters "still hunt," an increasing number of them were pursuing the prolific rabbit with packs of beagles. In

Successful rabbit hunter in the L'Anguille River bottoms near Pine Tree, 1987

In Arkansas, rabbits rank only behind deer and squirrels in terms of popularity.

Keith Sutton.

fact, the commission learned by a survey in 1984–85 that rabbits provided the third most popular form of hunting in the state. There was good news for rabbit hunters in 1987 when biologists reported rabbits were more numerous than usual, especially in central, west, and north-central Arkansas. They attributed this to "several mild winters followed by favorable weather during the early breeding season" and cautioned that the key to quality rabbit hunting was good habitat, which was decreasing every year.

SQUIRRELS

As noted previously, squirrel hunting has been regulated in Arkansas since 1917 when a bag limit of fifteen was set; it was reduced to eight in 1927. No other game in the state was subject to as much illegal hunting as the squirrel. Many persons who lived adjacent to squirrel woods felt they had the right to kill a "mess" of squirrels. Unless their habitat is destroyed, as it was rapidly in the 1940s and 1950s and still is through extensive timber clear-cutting in southwest Arkansas, squirrels have the reproductive capacity to withstand a heavy annual kill and still furnish good hunting from year to year (Holder 1951).

Arkansas has relied on Louisiana in the past for squirrel hunting information. In that state, as in Arkansas, there is usually some controversy among hunters about opening dates for the squirrel season. Research reveals that the squirrel population peaks in October, when squirrels are easier game than at any other time of the year. Even so, game management hunts in Louisiana from 1954 through 1965 showed that hunters averaged killing only about two squirrels per effort ("The Truth about Squirrel Hunting Success," 1968).

There appears to be little the state can do to increase squirrel populations except on lands devoted to wildlife. In these areas, timber can be managed to benefit squirrels by leaving den and mature mast-producing trees (Holder 1951). The 1984–85 survey showed squirrels were the most popular small game in Arkansas. More than half the state's 316,000 licensed hunters participated in the 1984–85 fall squirrel season and more than 50,000 took part in the spring season, making them the second largest hunting group behind deer hunters. Arkansas hunters take 3 million to 3.7 million squirrels annually. No records are kept to compare gray and fox squirrel harvests.

In 1987, wildlife biologists said Arkansans could look forward to the best squirrel hunting in some time because there had been "excellent mast crops for the past three years," which had caused the number of squirrels to increase steadily. There was a mast failure in 1989, however, and the squirrel population fell with it but rebounded by 1992. Because squirrels depend heavily on acorns, walnuts, hickory nuts, and other tree-produced foods, biologists reminded hunters to look for hardwood and mixed pine-hardwood stands, particularly those with red oaks.

Squirrel hunter and hound, circa 1945

A variety of treeing dog breeds are used to hunt fox and gray squirrels in Arkansas woodlands.

Arkansas Game and Fish Commission.

Furbearers

Professional trapping in Arkansas rapidly is becoming a lost art, and the days of the old-time trappers are numbered, Bill Apple opined in his December 5, 1954, wildlife column. His statement was based on the Game and Fish Commission's release of a report revealing the 1953–54 fur catch was only about half as valuable as that of the previous year. Only 120,776 pelts were marketed in the state for a total price of $312,073. The previous year saw more than double that number of pelts—242,350—marketed for a total of $574,222.

Apple attributed the sharp decline to the steady drop in all fur prices. Twenty years ago, he recalled, the business of trapping and dealing in furs was "highly profitable" both for the professional and the amateur. Almost any type of fur would bring a minimum of $7.50 a pelt, and even the lowly opossum was worth trapping. Quality mink pelts would bring up to $70.00 or $80.00 each, and a good, thick-furred raccoon would bring $15.00 to $20.00.

A few professional trappers were still around, Apple noted, but mostly the activity was in the hands of men and farm boys who wanted to pick up a little extra money. And they concentrated their efforts on the mink, which was about the only remaining furbearer with any commercial value. He explained:

> More than 90 percent of the value of the past season's fur catch was brought by mink pelts, which were marketed for $258,066 compared to the total market value of $312,073 for the entire fur catch of the year. The total number of mink pelts was 19,116, taken largely in the Delta along the Mississippi River.
>
> There were more raccoon pelts—51,508—brought to market than any other

W. H. Graves, Arkansas fur dealer, circa 1890

Animal hides and furs have been important items of trade in the state since the late seventeenth century.

Arkansas Game and Fish Commission.

single fur, but the total value . . . at 75 cents each was only $38,631. There were also a large number of possum hides marketed, 37,192, but at the cheap price of 15 cents a pelt, their total value was only a little more than $5,000. The only other major fur producing animals in Arkansas are the skunk and the muskrat, both of which are valued at less than one dollar each for their hides. Other fur bearers which play a minor role in the annual fur catch are the grey and red fox, civet cat, weasel and bobcat.

Mink pelts on stretchers, 1950

In 1954, 19,116 mink worth $258,066 were trapped in Arkansas. The value of the entire fur catch that year was $312,073.

Arkansas Game and Fish Commission.

RACCOON

In a column on December 16, 1956, Apple stated that the raccoon "has made a comeback unparalleled by any other game animal in the state, with the possible exception of the deer herd."

The 1930s were lean years for the raccoon because many of its den trees were cut and it was hunted for food and fur value more than for sport during the Great Depression.

The commission decided something had to be done to salvage the raccoon, Apple recalled, and members went to work on a restocking and education program. Sportsman groups got together and agreed "man to man" that they would not cut down the trees. "With this aid, Mister Raccoon made his comeback," Apple said.

Although raccoon skins did not have much commercial value, the interest in hunting them as sport increased through the years, Apple said. Trapping has never died in Arkansas but is limited today to a hardy group of mostly men and teen-aged boys who continue to do it for pleasure and to supplement their incomes. Wild pelt sales earned Arkansas trappers between $2 million and $3 million annually in the 1980s, with the most valuable and most plentiful being the raccoon. Raccoons favor densely wooded areas and rarely venture far from water, even in big cities like Little Rock. The Game and Fish Commission's veteran nuisance animal control officer, Rocky Lynch, said in 1992 that raccoons were the animals about which he received the most calls from populous central Arkansas.

Sustained campaigns by animal rights groups against humans wearing furs have contributed to recent depressed market prices, particularly in Europe. Whereas Arkansas trappers earned $1.7 million for their pelts in 1985–86, only $107,500.00 came in for 1992–93, averaging $3.00 for each of 18,110 raccoons, $.65 each for 2,630 opossums, ad $12.58 for each of 4,176 mink.

Because raccoons had become so plentiful by 1994, the Game and Fish Commission extended the hunting season on them for more than a month into the spring. Northeastern states began reporting problems with rabies in raccoons. Arkansas had no similar problems, but the Game and Fish Commission bowed to concerns of the state Health Department and wildlife disease agencies in 1996 and banned the importation and exportation of the animals without commission approval.

BEAVER

The beaver is another furbearer that has survived virtual extermination in Arkansas, much to the dismay of many of today's farmers and county judges. The Amendment 35 Game and Fish Commission should not be condemned or credited for the return of beavers to the state. The old commission brought a few beavers into Arkansas in the early 1940s. Then, on February 14, 1944, it gave this valentine to the state:

> On presentation of a project to the Commission whereby the department will engage in beaver restoration work . . . on a motion by Mr. Lawton . . . the

In the mid-1940s, the Game and Fish Commission obtained live beavers from other states and stocked them in Arkansas in an attempt to stimulate the trapping industry. When trapping dropped to an all-time low, the beaver quickly became overpopulated and a nuisance animal.

Arkansas Game and Fish Commission.

> Commission authorized the purchase of six beaver traps at $20 from the National Live Trap Co. at Tomahawk, Wis., and other necessary equipment in order that the Commission might work with the federal Fish and Wildlife Service in securing primary brood stock for several areas to be designated.

The door opened. Eight beavers were brought from Alabama and released in Calhoun and Conway Counties. In 1945, more beaver were brought from Alabama and released in four county refuges. In addition, ten beavers were brought from Mississippi in 1945. In all, seventy-seven beavers were released in the state. Most of the restockings of southern beavers were successful (Pledger 1970).

A motivating factor at the time was the lucrative fur market. In the mid-1940s, a beaver pelt brought twenty dollars. By 1949, as was shown in part 2, the bottom dropped out of the fur markets and the beaver was no longer trapped heavily. In 1974–75, commission records show that only 889 beavers had been trapped by the 3,710 trappers in the state. The average price in 1976 for a beaver pelt was $4.93.

Beaver in live trap

Today the beaver is widely regarded as a nuisance in Arkansas. Even so, the beaver is a conservationist in its own right. Their ponds provide homes for many wildlife species including waterfowl, muskrats, and otters (Pledger 1976).

Conservationist Harold Alexander has a naturalist's view of the beaver. In *Arkansas Game and Fish* (spring 1978), he said the beaver, like most of us, was

> neither all good or all bad. In the lowlands, for example, he builds small dams which form highly desirable wetland habitats for ducks, raccoons and other species which live and feed in habitats of this type, and which habitats are disappearing at a rapid rate. In the hill country, he constructs these dams of sticks and mud across streams in such a way that they create deep pools which retain water through the drought periods, act as cooling and filter basins, and continuously release water downstream to maintain a viable stream system.

Beavers being released

> In other words, the beaver is neither "friend or foe" but a valuable furbearing and often beneficial mammal which should, like deer, ducks or raccoons, be managed in relation to the variable circumstances in which they may be found.

About every five or six years, county judges and farmers descend on state legislators, demanding that something be done about the beavers that cause water to back up and damage crops, roads, and bridges. The latest in this cycle occurred in 1991–92, when talk surfaced about establishing a state bounty on beaver pelts.

Rocky Lynch, the commission's animal control officer, warned the lawmakers against this, explaining that the state would end up paying for beavers imported from adjoining states. Trappers out for fast dollars would go to hot spots, snare as many beavers as possible, and leave. Some beaver would remain, and so would the problem. Lynch said only sustained trapping by those who know what they are doing would bring the situation under control.

The beaver problem became so severe in Phillips County that quorum court member Clark Hall described to the legislative Joint Interim Committee on Agriculture and Economic Development how his area had initiated a control program in 1989. The quorum court appropriated five thousand dollar in seed money for the program that was matched by fifteen thousand dollars raised locally. Phillips County was divided into six districts with a local trapper hired to work each: "I'm running a bounty system . . . they supply their own traps . . . and I'm paying $10 a tail," Hall said.

Harold McAlpine, president of the Arkansas Trappers Association, attributed some of the problem to the fact that more and more landowners were restricting access to their property. He urged county judges and landowners to establish better relations with local trappers and open their lands. A central referral center for the names of trappers also would be helpful, he advised. Eventually, the state Department of Finance and Administration was given the legal responsibility of paying five dollars a beaver tail to supplement what counties paid in their programs.

RIVER OTTER

The Game and Fish Commission always has pointed out that "Not everything a beaver does is damaging." For example, dams that flood valuable property also create aquatic habitat for an amazingly diverse variety of wildlife. Otters, wood ducks, white-tailed deer, largemouth bass, and snapping turtles are just a few examples. The otter, in particular, owes its comeback in Arkansas to the beaver.

Overtrapping, primarily to meet European fur demands, virtually wiped out the otter in Arkansas in the late nineteenth and early twentieth centuries. The otter finally was protected from trapping by the state legislature in 1915. The animal's problems were only beginning, however, because of widespread habitat destruction and pollution. Otters are at the top of the aquatic food chain and accumulate long-lived environmental pollutants from the foods they eat. Research biologists concluded that possibly the extensive use of DDT, heptachlor, dieldrin, and certain heavy metals during the 1950s and 1960s further inhibited reproduction or lowered the otter's ability to survive.

Beaver impoundments typically have high fish populations, contain much cover, and have low turbidity levels—ideal otter habitat. The otter once again is occurring in healthy numbers in Arkansas because of the beaver, trapping regulations, and bans on many dangerous pollutants.

Furriers consider otter to be the diamond of their business because its pelt is dense and durable. Arkansas otter harvests ranged from 25 to 749 pelts annually over the twenty years that ended in 1986. The value of the pelts has fluctuated widely. Average prices paid for individual pelts during the twenty-year period ranged from a low of $12 in 1967 to a high of nearly $44 in 1978.

MUSKRAT

Apple was glad to see the raccoon return. He had little if anything to say about the beavers, which were just beginning their comeback in his heyday as a column writer. But he definitely did not like the muskrat and did not hesitate to say so in his October 28, 1958 column: "The muskrat is classified as one of America's top fur bearers, but in my book and a lot of others, he is an expensive squatter." He explained,

> The muskrat's habit of denning causes most of the damage. Hiding the entrance to his den under water, he burrows upward where he hollows out his home. Branching from there, he munches on tender roots and vegetation for much of his food. If he burrows through the levee, then that means an immediate leak and a levee to repair.
>
> Like other members of the rodent family, the muskrat is a prolific breeder. Since I definitely do not like muskrats because of their destructive habits, in turn, I would form a definite like for the mink that keeps their population in balance.

Muskrat capture

Muskrats are common throughout Arkansas today, causing serious problems by their burrowing in levees of lakes and ponds. In the 1940s, however, they were absent from the Gulf Coastal Plain region. In 1942 and 1943, many were caught in live traps at Big Lake National Wildlife Refuge and released in areas formerly uninhabited by muskrats.

Arkansas Game and Fish Commission.

Even the Game and Fish Commission has conceded that muskrats can be "serious pests" with damages to Arkansas rice growers perhaps reaching $1 million a year. They can be controlled somewhat by trapping, and even though their pelts do not always fetch high prices, the ease with which they can be caught and processed has kept them popular with trappers.

FOXES

> If you want to match wits with an animal near the top of the intellectual plane, arrange an introduction to the red fox. Whether you take him by trap or chase, you'll agree he's worthy of all the fine adjectives used to describe him.
>
> A cohort of the red fox, the gray fox, is not nearly so valuable either for sport or pelting. The gray fox's habits are much the same as the red's except when pursued by hounds. He is quick to find a tree. Thus the sport of the chase is lost and only the value of his pelt is left.

So wrote Bill Apple on December 16, 1956.

It would have been tough for many years to arrange an introduction to a red fox in Arkansas because there were so few left, and they were protected by a statewide closed season. The role of the coyote in the red fox's decline was discussed in part 2. Biologists also believe the clearing of timberlands and the effect of livestock grazing had reduced the fox's favorite foods—mice and rats. In an effort to reverse the decline, the Game and Fish Commission gave the Arkansas State Fox and Wolf Hunters' Association a permit to bring twelve pen-raised red foxes from outside the state if they could be used in a research project. The Camp Robinson Quail Restoration Area adjacent to Lake Conway was chosen for the study. Six foxes were outfitted with small radio transmitters and released on the area. Five were released with identifying ear tags. Five of the eleven foxes in the study were killed by various causes during the first five weeks after release, indicating the survival rate of young pen-reared foxes is rather low (Pledger 1975). In the 1990s, hunters who chased the red fox with dogs but did not kill them, trappers, and other outdoorsmen began reporting seeing more and more of the colorful, crafty furbearers. The Game and Fish Commission took a cautious step in 1996 to reopen the season by allowing hunters and trappers to obtain a tag for each red fox, up to six, they planned to take.

Once common in fields and pastures, the gray fox has given way in Arkansas to the coyote and now inhabits heavier cover where it can climb trees to escape its larger enemy.

The Arkansas Health Department asked the Game and Fish Commission in 1995 to ban the importation of live foxes and coyotes to guard against the spread of canine distemper, rabies, and the *Echinococcus multilocularis* tapeworm, which can be transmitted to humans and cause death.

Coursing pen owners, primarily in north-central Arkansas, howled in protest, saying their businesses could not survive without the importation of foxes. Coursing pens are large fenced tracts where sportsmen pay to follow their dogs as they chase foxes and coyotes. A compromise was reached under which those who import must obtain a permit, and the animals they bring into the state must have health papers.

Red fox

The red fox has an intersting history in Arkansas. Living at the southern edge of its range, it prefers more open habitat than the gray fox. Its habitat expanded from the prairies to include small farms when settlers cleared the forests. But the clearing of the forests also allowed the coyote to expand its range. Competition between the two has favored the coyote. Never a common sight in Arkansas, the red fox is most readily found in the northern Delta and the Ozarks.

Len Rue Jr.

Any animal that dies in transit or captivity also must be submitted to the state Livestock and Poultry Commission for laboratory tests within seventy-two hours.

COYOTE

Coyotes, which began expanding their habitat into Arkansas in the 1940s and reached the Mississippi River by 1964, now occur in all counties; one coyote per two to four square miles is a typical density except around poultry dumps or other food sources where they may number two to three per square mile. They are thriving in Arkansas, and no formal management has been deemed necessary, though in 1994 the season for hunting them was extended a month. From 1976 through 1986, three to fifteen thousand Arkansas coyotes were harvested annually, most in the Ozark Mountains and the Arkansas River Valley. They are classified formally as furbearers.

Proponents of coyotes are pleased by the expansion: wildlife watchers want to see and hear them; varmit hunters enjoy calling coyotes or running them with dogs, and trappers make extra money from their pelts. But many livestock producers and melon growers demand their numbers be reduced. The commission has an extension trapping program for farmers with livestock or melon losses to coyotes; it teaches farmers to deal with depredation problems caused by the coyote.

OTHER FURBEARERS

The status of other Arkansas furbearers in the 1980s and 1990s was reported:

> The opossum remains one of Arkansas's most abundant mammals, largely because of its adaptability.
>
> The nutria, an uninvited 1938 import from Louisiana, is another pest for farmers because it is well-established in eastern and southern parts of Arkansas where it digs into stream banks and levees and damages crops.
>
> The long-tailed weasel is considered rare, but its highly secretive nature makes it seem more scarce than it really is.
>
> The striped skunk, common statewide, lost its value as a furbearer when 1951

Coyote howling

The coyote is highly esteemed by predator hunters and by Arkansas houndsmen who enjoy pursuing them with dogs.

Leonard Lee Rue III.

truth-in-labeling laws forbid it from being marketed any longer as "American sable." The striped skunk is the predominant carrier of rabies in the state.

Spotted skunks occur throughout Arkansas but are not common.

The badger remains the most rare furbearer in Arkansas with fewer than twelve pelts being taken annually. Badgers occur primarily in the northwestern half of the Arkansas River Valley.

The mink remains one of Arkansas's principal furbearers and is common along brushy waterways throughout the state.

Contrary to public perception, all furbearers are not taken by trappers. Many are taken, especially during the furbearing season, by dog hunters who pursue running game such as raccoons and foxes. During the 1977–78 season, fur-dealer records show that 65 percent of raccoons and 33 percent of gray foxes were taken by hunters. Some hunters who use dogs to chase furbearers seldom kill the game once it is treed. Organized field trials—mostly for chase of raccoons and foxes—have evolved into rather large events, often attracting more than two hundred dogs and three hundred-plus sportsmen of all ages. Dogs often are bred specifically for field trials, and a good running and/or bench show dog can be worth hundreds, sometimes even thousands, of dollars (Pledger 1979).

Coon hunting hounds

Leonard Lee Rue III.

Predators

[P]redatory animals have a place in nature. The only basis for a desire to exterminate them is prejudice or ignorance. Not a few of today's wildlife problems are due to the utter lack of understanding and appreciation of our predatory animals. Some years ago a German literary scholar now living in this country remarked: "I'm greatly interested in snakes. The treatment they receive in rural America disturbs me. I know from experience in Europe that they have a place, and you get into trouble when you fail to recognize that fact."

John D. Black, Ph.D
The Management and Conservation of Biological Resources

Wildlife flourishes only on land that is healthy and diverse. Game management people learned that it is sound economics to improve an animal's habitat rather than to try to exterminate its predators. The idea is not to raise the largest possible deer herd, but to confine the size of the herd to the land's "carrying capacity.". . . If predators and regulated hunting do not provide the necessary checks on that species' population, then starvation takes its toll.

Frank Graham Jr.
Man's Dominion

Anyone who has studied wildlife populations in various habitats will have observed that practically everywhere animals live in communities complete with herbivore and carnivore components. This is true even in our most artificialized environments. In fact, it may seriously be questioned whether an animal community can exist for long in anything but relative balance.

Durward L. Allen
Our Wildlife Legacy

> Predators play an important role in maintaining the balance of nature. This is a proven fact . . . The trapping of coyotes, wolves, and bobcats by professionals is, poor thinking . . . I am making a plea to all of you to think of not only the economic value of the predator but also think of them as having an aesthetic value which should be passed on as a heritage to those who come after us.
>
> Dr. Howard K. Suzuki
> *The Role of Predators*

In light of these views, the Game and Fish Commission was a victim and tool of both politics and ignorance in the early 1950s when it carried out a predator control program mandated under the last law of its kind by the Arkansas General Assembly. From July 1, 1949, through February 28, 1950, commission-hired professional trappers reported they had snared 149 wolves and 42 bobcats. In July 1953, the trappers reported taking 300 wolves and 99 bobcats the previous year. This only made up a fraction of what agencies of the federal government were doing to wipe out predators, particularly in the West.

RED WOLF

The wolf that frightened and fascinated settlers in Arkansas was *Canis niger*, the red wolf, a species indigenous to North America that ranged from central Texas to Florida and north to the Ohio River Valley. Analyses of red wolf skulls collected in the south-central part of its range, including south Arkansas, show almost every red wolf was considerably larger than the largest coyote, but the majority were smaller than most gray wolves. Females usually weighed from forty to sixty pounds and males from fifty to seventy-five pounds. The red wolf was more slender and had relatively longer legs than the gray (timber) wolf that ranged the northern United States and Canada. These features are believed to have evolved in the red wolf for long-distance running and to help it maneuver in the coastal prairies and river bottom swamps where it was particularly abundant.

In color, the red wolf is considered one of the most beautifully marked of all in the wolf family. The body hairs can be reddish, interspersed with black hairs. Some people say it appears to have "jumped out of a pot of red barnroof paint." This is an exaggeration if not patently incorrect. According to expert Dr. Ronald M. Nowak, the color that gives the red wolf its name actually is not a common occurrence, and most specimens have a typical gray-brown or tawny coat with black hairs interspersed.

Little is known about the ecology, reproduction, and behavior of the red wolf because no significant studies were made when the species was abundant, and later efforts were limited by the problem of locating pure red wolf populations.

Hunters with red wolf killed near Red Star, Arkansas, circa 1920

Hounds trained to chase and bay wolves were common during the first half of the twentieth century. Large groups of men often met to participate in hunts aimed at exterminating these predators.

Arkansas Game and Fish Commission.

Wolf trapper

In the late 1940s and early 1950s, the Game and Fish Commission hired professional trappers to carry out a predator control program mandated by the Arkansas General Assembly. This unidentified professional poses with a young red wolf captured in the Pulaski County Game Refuge.

Arkansas Game and Fish Commission.

In addition to the state and federal government's trapping efforts until midcentury, the red wolf took a major environmental hit in the 1920s when the great forests that once covered the south-central United States were wiped out. Remnant populations could be found as late as the 1950s along the Gulf Coast and in bottomland forests in northeastern Louisiana and eastern Texas.

Habitat changes also brought a breakdown of behavioral patterns and ecological isolation. When deprived of a mate of its own species, a red wolf apparently will breed with coyotes and other canids, producing widespread hybridization.

The rising tide of environmental awareness in the 1960s led the U.S. Fish and Wildlife Service to take note of what was happening to the red wolf, and it was placed on the agency's rare and endangered species list in January 1965. Federal predator-control operations in the red wolf's range were curtailed a year later and active investigation and protective efforts started in March 1968.

Dr. Douglas Pimlot of Toronto University, a world authority on wolves, came to Arkansas in 1964 to evaluate the status of the red wolf. One study method Dr. Pimlot used was to travel areas where the red wolf had ranged and play recordings of northern wolf howls similar in tone and volume to those of the indigenous wolves. He and his aides listened for responses. They got only two in Arkansas—one from a wolf in the Black Mountain area of Franklin County and another in southwest Arkansas. Through these surveys and other techniques, it became increasingly evident the red wolf had disappeared from the woodlands where twenty years earlier their spine-tingling howls often could be heard in the night. Even so, the Arkansas Game and Fish Commission did not act until 1969 to place red wolves in protected status.

To preserve the genetic purity of the red wolf and save it from extinction, the U.S. Fish and Wildlife Service began rounding up the surviving wild animals in the mid- and late-1970s. Of the more than four hundred canids captured, only seventeen were found to be purebred red wolves. The species was declared extinct in the wild, and a captive breeding project was set up at Point Defiance Zoo in Tacoma, Washington. The program was a success, and efforts have been made since to locate suitable sites for red wolves to be released in the wild. One such site was the Alligator River National Wildlife Refuge, a wilderness in North Carolina, where red wolves bred in captivity were released in 1987. The project has been controversial with some residents, and even some professionals question whether the subspecies of red wolf brought to the refuge ever really existed in North Carolina. Another relocation site for the red wolf was Horn Island off the coast of Mississippi, which received two adult wolves from Alligator River in 1989. Two pups were born to relocated parents in 1990 at St. Vincent National Wildlife Refuge, which is on an island off Florida's Gulf Coast.

In Arkansas, the Game and Fish Commission and the University of Arkansas Zoology Department conducted a four-year study in the 1970s to identify wild canids.

Through this project it was determined Arkansas's existing wild canid population consisted of about 73 percent coyotes, 13 percent coyote-dog hybrids, three percent wild dogs, 1 percent red wolves, and 1 percent red wolf-dog hybrids. Thus the state's wild canid population is composed of several elements functioning as a single interbreeding unit. Though coyotes predominate, "occasionally some individuals are seen with all the physical traits of the red wolf, coyote-dog crosses that result in odd color patterns, and individuals impossible to identify without careful examination." One of the researchers said that as the red wolf declined, the coyote population expanded and gradually replaced it as the dominant canid (Gipson 1974).

Discarded poultry is the most important food during all seasons for many of Arkansas's wild canids, except in the fall when persimmons become their number one food item. A significant number of turkeys are killed in rearing pens by wild canids. On several farms, open dumps for dead turkeys were observed near raising pens. The farmers were feeding dead turkeys to coyotes on one side of the fence and expecting them to leave the live turkeys on the other side alone. Coyotes do not understand such logic (Gipson 1974).

The commission receives more complaints about calf predation than for any other type of wild canid damage. The examination of fifty-eight wild canids, trapped in response to complaints about calf losses, showed only six—12 percent—actually had ingested calf tissue.

BOBCAT

Contrary to what some antihunters have claimed, the bobcat is not endangered today. They are reported to be holding their own and even increasing in many areas. Bobcat pelts vary in value but often command the highest price in Arkansas.

The bobcat fared far better than the red wolf. In fact, "the bobcat, at least in Arkansas, is doing well," according to Dr. Gary A. Heidt, a biology professor at the

Bobcat

During the 1970s, a ban on the importation of various species of cats from abroad resulted in a greatly increased demand for bobcat pelts. For several years, individual pelts often fetched one hundred dollars or more on the fur market.

Keith Sutton.

Dead panther

This 118-pound male panther was killed near Lake Dardanelle in Logan County in November 1975. Some speculate that this was the last wild lion shot in Arkansas. Biologists are unsure, however, if this is a wild native cat or a pet that escaped or was released.

Arkansas Game and Fish Commission.

University of Arkansas at Little Rock who participated in a $170,000 federally funded research project in 1984 by three universities and the commission. A census was not the intent of the study, so no one is sure how many bobcats are in Arkansas. The researchers did find, however, an average of one bobcat for every 3.5 square miles in the Ouachita Mountains. That ratio is not constant throughout Arkansas, but if it were, this would mean perhaps fifteen thousand bobcats are in the state.

The eight bobcats radio monitored during the project disappeared or were killed within three months, revealing the fragile existence of the animals that once were considered a nuisance but have been protected as a furbearer since the mid-1970s. Dr. Heidt said Arkansas cannot be complacent about bobcats: "As we lose more wetlands and hardwoods, it can't do anything but hurt them" because these are bobcats' habitats.

FLORIDA PANTHER

Florida panthers (mountain lions) are not legal game in Arkansas and are protected by a permanently closed season. They also are protected by federal and state endangered species laws, with heavy fines possible for those who injure or kill one of the big cats—if they ever really encounter one.

In 1988, the Game and Fish Commission hired Rancher's Supply of Alpine, Texas, to conduct in-depth field investigations for physical evidence of native panther populations in Arkansas. Famed lion tracker Roy McBride, the company's owner, began his field work in August 1988, searching for recognizable mountain lion sign such as tracks, urine markers, scats, and kills. In his first progress report to the commission, he reported covering 1,900 miles in the Ozark National Forest and more than 450 miles in the Mount Magazine area without finding any evidence of mountain lions.

McBride said, "There were no truly remote, inaccessible areas within the forest that could harbor an undetected population of mountain lions." Further, extensive use of the forest by canoeists, trappers, hikers, loggers, and all-terrain vehicle recreationists invariably would lead to sightings of several wild mountain lions annually if they were present, McBride said, and numerous highways interspersed throughout the Ozarks carry a high volume of traffic that would cause the discovery of lions as road kills. McBride concluded that "the area surveyed during this preliminary investigation did not have a resident, breeding population of wild mountain lions."

McBride turned his attention to the Ouachita National Forest and other areas with suitable size and wilderness characteristics necessary to support viable mountain lion populations, but his search was unsuccessful in those places as well.

Turtles

A nongame regulation adopted by the commission in January 1987 caused considerable confusion among turtle fanciers. The regulation made it illegal to take any wildlife unless an open season had been established. The only season for turtles was on aquatic turtles. Aquatic turtles are those living primarily in water and include all species seen swimming or basking on a log in a lake or stream—snapping turtles, mud turtles, musk turtles, map turtles, sliders, red-eared turtles, chicken turtles, and softshell turtles. All turtles in Arkansas, except box turtles, are considered aquatic.

Selling box turtles, as some were doing for pets or for turning into knicknacks, was made illegal, but having up to six as pets was made permissible as long as they were not ornate box turtles. The ornate box turtle is a creature of open, treeless areas and rarely is found outside the few prairie grassland areas left in the state. Its numbers have declined throughout its range because the ever-increasing amount of land turned to agriculture and the ever-expanding highway system. The three-toed box turtle is usually found in mature forests and thickets. They are more common.

The state's commercial turtlers, who harvest aquatic turtles, were upset with the commission again in 1990 when the panel, responding to concerns by wildlife officers that turtle nets had the capability of catching some fish, changed regulations once again. The new rule said commercial turtling was restricted to waters open to regular commercial fishing and required turtlers to obtain a commercial tackle license for each turtle net and turtle box they used. Kenneth Dowdy of Rector and James Stout of Corning protested vigorously at a subsequent meeting, pointing out that the new regulations conflicted with the published 1990 fishing regulations guide. The commission rescinded the regulation for the remainder of the 1990 fishing season and left the solution to be worked out in the future.

Wildlife biologists went to the Game and Fish Commission in October 1993 and asked for an emergency ban on the taking of alligator snapping turtles from public waters because they were alarmed about what they saw as the decline of the species. The agency granted the ban and asked for a survey.

Alligator snapper

The alligator snapper is one of the world's largest freshwater turtles. This specimen from the St. Francis River in Cross County weighed more than 130 pounds. The largest on record is a wild-caught specimen weighing 316 pounds.

Keith Sutton.

Prairie County wildlife officer Ernest Ingle said Louisiana turtle trappers moved into Arkansas in the mid-1980s and "wiped out the alligator snappers in several streams such as LaGrue Bayou and tributaries of Peckerwood Lake. Other streams such as Bayou Bartholomew, Bayou Two Prairie and Wattensaw Bayou were virtually clean of snappers of reproductive size. When it became financially infeasible to trap in eastern and southern Arkansas, the trappers concentrated their efforts in north-central and northeast Arkansas, depleting such streams as Salado Creek." Habitat loss because of river channelization and damming has contributed to the dwindling population as well.

Alligator snappers closely resemble common snapping turtles. The most obvious difference is the triple keel of pointed scales on an alligator snapper's back, which is absent on common snappers. The alligator snapper's most distinctive feature, however, is its size. For example, John Williamson of Elaine found two ensnared in his trammel nets on the White River in 1987. One weighed 116 pounds and its companion was a 131-pounder. The flesh of alligator snappers is considered a delicacy, often fetching more than ten dollars a pound, but Williamson freed the ones he caught.

The biting ability of alligator snappers in legendary. Claims of alligator snappers biting through broom handles have been verified by tests, but even the scissor-like shearing surfaces of an alligator snapper's jaws cannot sever a man's leg, as is sometimes rumored. Outdoor writer Jim Low adds, however, "there are folks walking around with fewer than 10 fingers as a result of alligator snapper encounters."

In January 1995, Game and Fish Commission biologists reported to the panel that their survey had found only 425 alligator snappers located in forty-nine counties. Snappers previously had been unknown in thirty-six of those counties. The ban continues.

NONGAME WILDLIFE

> Like a piece of fabric that is made up of many strands of thread, life on earth is made up of many living things, each closely related to the others. Remove a strand of thread from a piece of fabric, and the cloth will be weakened. Remove many strands, and the cloth will fall apart. In the same way, the loss of one or two species seems to do little harm, but the loss of many can have serious consequences.
>
> Sunni Bloyd
> *Endangered Species*

The environmental movement of the 1970s brought unprecedented attention to and appreciation for nongame wildlife by the public. The Arkansas Game and Fish Commission came under increasing pressure through the years from such groups as the Audubon Society and the Arkansas Conservation Coalition to protect and preserve nongame species. In 1979, the commission proclaimed its biologists and wildlife officers were "fully aware" that Amendment 35 to the state constitution made the agency responsible for controlling, managing, restoring, and conserving Arkansas's wildlife resources and that the broad charge encompassed nongame species. Not until early 1987, however, did the commission act to clear up what it called "public misconceptions" about its authority over all Arkansas wildlife by adopting regulations governing the taking and commercialization of nongame in the state.

The long-standing restriction against the buying, selling, bartering, or trading of game animals except by those holding commercial, trapping, breeding, and other special permits, was enlarged to cover nongame species. The rule also made it unlawful to

> Take from the state and sell or offer for sale as fishing bait any wildlife species or portions thereof other than insects, freshwater shrimp, worms, frogs, minnows, shad, mussels and crayfish
>
> Take or try to take by any method other than by hands and possess alive any wildlife species other than two (2) each of the animals specified [bobcats, coyotes, deer, gray foxes, opossums, quail, rabbits, raccoons, and squirrels]. The sale, purchase, barter or transporting of the animals from the state or their progeny was prohibited.

The regulations extended the "hands off" list to nongame birds, bats, hellbenders, ornate box turtles, cave-dwelling species, and those animals listed as endangered species. Zoos were excluded, as was wildlife exchanged by the commission with other states.

Those who were collecting and selling certain nongame species mounted a major effort to get the new regulation repealed in May 1987. The commission also fielded complaints from those who wanted to be allowed to take a few more of such common species as the ordinary three-toed box turtle for pets. The only change the commission made, on a motion by member Perry Mikles of Booneville, was to increase the number of three-toed box turtles that could be taken as pets from two to six.

The federal Endangered Species Act of 1973 was the first law to extend protection to all classes of animals, including invertebrates, and was the first to recognize threatened as well as endangered species, thus establishing federal authority for protection before the danger of extinction became grave. It also provided for export controls and for listing and conservation of plants, both of which had not been included in previous legislation.

The law provided "a means whereby the ecosystems upon which endangered species depend may be conserved, protected and restored" by designating "critical habitat" to be protected.

The commission's 1987 nongame regulation identified these species as "endangered or threatened" in Arkansas: gray bat, Indiana bat, Ozark big-eared bat, Florida panther, bald eagle, Arctic peregrine falcon, interior least tern, red-cockaded woodpecker, American alligator, Ozark cavefish, leopard darter, Curtis' pearly mussel, pink mucket mussel, fat pocketbook pearly mussel, ivory-billed woodpecker, and Bachman's warbler. In 1991, four endangered mussels native to Arkansas were added to the list. They were the turgid blossom mussel, speckled pocketbook mussel, Arkansas fatmucket mussel, and the Ouachita rock pocketbook. Also added were the Magazine Mountain shagreen land snail, the pallid sturgeon, and the American burying beetle. A new cave crayfish species was found in Stone County in 1992, joining two others in northern Arkansas that are endangered.

Some endangered species, such as the American alligator and bald eagle, are making comebacks, while others are merely holding their own or are sliding even closer toward extinction.

The following includes selected other nongame species common in Arkansas will be described.

Invertebrates

FAT POCKETBOOK PEARLY MUSSEL

Although once found in the White River, the only verified remaining populations of this species occur in the St. Francis River in Cross and St. Francis Counties. Dredging by the Corps of Engineers was in progress on the St. Francis River when the discovery of specimens of this endangered species in the spoil area forced work to stop, pending consultation with the Fish and Wildlife Service. Dredging resumed in the late 1980s—more than a dozen years later—after a plan designed to protect the mussels was worked out. Government spokesmen said studies have shown the dredging has not harmed the mussels.

OTHER MUSSELS

Several mussel buyers approached the commission in 1991 to increase the minimum size for mussels to be harvested because they said the resource was being

Biologist John Harris, an employee of the Arkansas Highway Commission, surfaces after a dive in the Saline River near Benton to search for endangered mussels. Seven species of Arkansas mussels were on the federal list of endangered species in 1997, including the fat pocketbook pearly mussel pictured. Presence of these protected species is sometimes cause for delaying construction of bridges and other highway projects.

Keith Sutton (left) *and Arkansas Game and Fish Commission* (below).

threatened. Enforcement Division chief Loren Hitchcock advised the commissioners that there had been "a tremendous increase" in recent mussel shell taker, buyer, and seller license sales, that the reporting requirement needed to be more stringent, that the minimum size needed to be increased, that the residency requirement to take mussels needed to be at least twelve months, and that license costs needed to be increased. The commission responded by taking these actions and setting the residency requirement at one year.

Above left, **Diver and assistant**

Above right, **Fat pocketbook pearly mussel**

Fishes

OZARK CAVEFISH

Only fourteen caves in the world are known to contain Ozark cavefish, an endangered species, and four of them are in Arkansas. A cave in Benton County contains the largest population, counted in 1983 at 97 and 1984 at 102.

Though probably never abundant, the cavefish is thought to be declining because of overcollection, groundwater pollution, and the decline of the gray bat, whose guano provided the caves' primary energy source and food supply.

Leopard darter

Ken Collins, U.S. Fish and Wildlife Service.

LEOPARD DARTER

In Arkansas, this species is known only in the Mountain Fork River and Cossatot River watersheds in the southwestern part of the state. Dam building on the Cossatot River and a chemical spill on the Mountain Fork destroyed much of the fish's upland river habitat, and it is classified as threatened.

Amphibians

TOADS AND FROGS

Twenty-one species of frogs and toads occur in Arkansas, compared to nearly twenty-seven hundred worldwide. Two species of spadefoot toads live in the state. Two subspecies of the eastern spadefoot occur in Arkansas. The eastern subspecies has been found in Chicot, Desha, Mississippi, Poinsett, Craighead, Lawrence, and Randolph Counties. Hunter's eastern spadefoot occurs in southwest Arkansas roughly along a line from the Missouri/Oklahoma border to Union County. (It may be a separate species.) An isolated population exists in Cleburne and White Counties. The plains spadefoot has been recorded in Arkansas only at Holla Bend National Wildlife Refuge in Pope County.

The most common toads in Arkansas are member of the Bufonidae family, which contains three species. All are distinguished by a large, swollen parotid gland above each eye that secretes chemicals called bufotoxins. Enough of this can cause nausea, irregular heartbeat, and even death, which explains why a dog or other predator that eats a toad is not apt to repeat the mistake.

There are eight species of treefrogs in Arkansas. Cope's gray treefrog, the

Northern spring peeper, the cricket frog, and the chorus frog are distributed statewide. The others and their locations are bird-voiced treefrog: Sevier and Little River counties near the juncture of the Arkansas, Texas, and Oklahoma state lines, along the Mississippi River, and in Pope, Conway, Perry and Saline Counties; green treefrog: the southwestern two-thirds of the state and in the Arkansas River Valley west to Oklahoma; and Strecker's chorus frog: only in Clay County.

Arkansas has only two species of narrowmouth frogs, which are in the Microhylidae family. The eastern narrowmouth frog is found statewide, and the Great Plains species is found only along Arkansas's western border from Polk to Crawford Counties.

Bullfrog

Arkansans harvest thousands of bullfrogs annually. The hind legs of these large amphibians are considered a delicacy.

Arkansas Game and Fish Commission.

Members of the Ranidae family are called true frogs, and there are six species in Arkansas. Those having statewide distribution are the bullfrog, the southern leopard frog, and the pickerel frog. The green frog has two subspecies in Arkansas—the bronze frog occurring in South Arkansas and the green frog in the northern half of the state. The southern crawfish frog is found in counties along the Oklahoma and Texas borders to northeast Louisiana, as well as along the Arkansas River from Oklahoma to the Mississippi River. The northern crawfish frog is found in Bradley, Drew, Chicot, and White Counties. Some authorities consider the western and eastern populations separate species.

The wood frog is perhaps the most interesting of the amphibians in Arkansas. It is the only North American frog to be found north of the Arctic Circle. There are two large, isolated populations in Arkansas, probably remnants from the last Ice Age. One is in Newton, Johnson, and Pope Counties, and the other is in Baxter, Izard, and Independence Counties.

Reptiles

SNAKES

There are forty-six varieties of snakes in Arkansas; of these, only six are poisonous—the copperhead, western cottonmouth, western pigmy rattlesnake, timber rattlesnake, western diamondback rattlesnake, and Texas coral snake. All except the coral snake are pit vipers, which refers to a sensory organ between and slightly below the eye and nostril on each side of the snake's head. The organ allows the snake to detect temperature differences in its environment, which aid in locating prey and directing lethal strikes on warm-blooded animals.

Rattlesnake

A 1977 visitor at Village Creek State Park near Wynne displays a timber rattlesnake found laying on the road at night. Six of the Arkansas's forty-six varieties of snakes are venomous.

Barry Milton.

AMERICAN ALLIGATOR

Alligator. Even the word sounds mean and ugly, but that did not stop the commission from restocking Arkansas with the reptile. Recognizing that alligators were threatened with extinction, primarily from loss of habitat through draining of swamps and other wetlands, the commission enacted a regulation in 1961 to protect the animal.

From 1972 until 1984, Arkansas biologists trapped subadult alligators in Louisiana and moved them to Arkansas. Many were planted in White River National Wildlife Refuge, but 80 percent went to farms whose owners thought they would make good beaver controllers. Louisiana was glad to lose the alligators because some of its parishes were knee-deep in them, to the extent that controlled hunts had been allowed.

By 1976, Arkansas was reported to have about 1,000 alligators. Grassy Lake near Hope was said to have the highest concentration. This conflicts with another biologist who said in 1985 there were an estimated 3,500 alligators in Arkansas in 1972 before 2,800 were brought in from Louisiana (James 1985). By 1987, it was reported forty-five Arkansas counties had alligators, including six where natural reproduction of the reptiles had not been documented previously. The alligator population in Arkansas seems to be stable and perhaps even growing slightly.

Some people harbor thoughts of raising them, and Henry Platt of Sevier County received Arkansas Alligator Farm Permit No. 1 in September 1992 from the Game

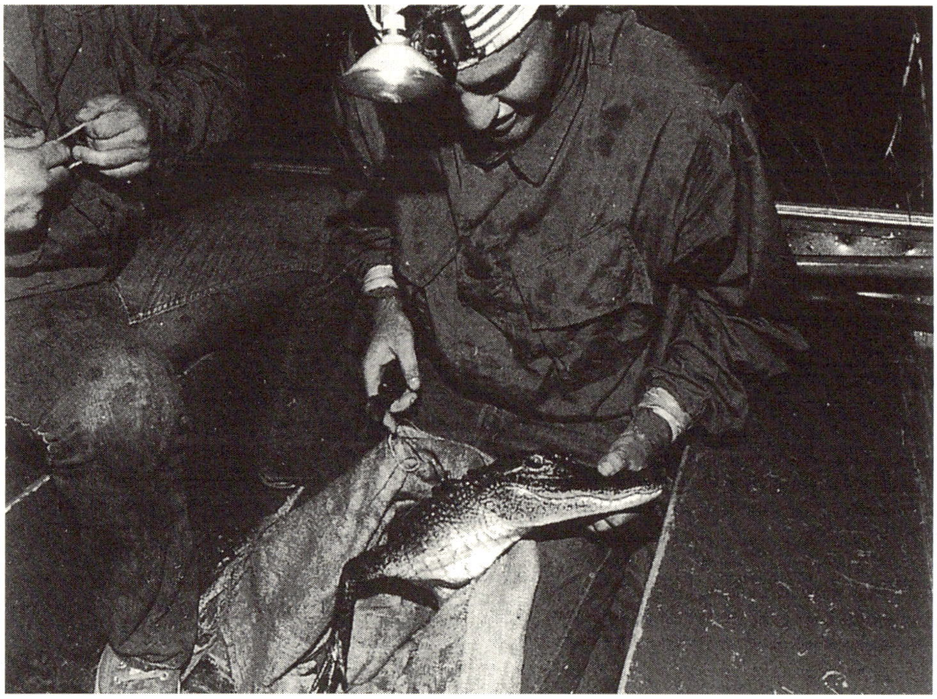

Alligator capture

A young alligator from Louisiana's Rockefeller National Wildlife Refuge is placed in a capture bag prior to transport to Arkansas. Almost three thousand Bayou State 'gators were released in Arkansas during restoration efforts in the 1970s and '80s.

Arkansas Game and Fish Commission.

and Fish Commission for this purpose. There is still concern for the species. Although it appears to be secure from extinction, the alligator is designated as "threatened, by similarity of appearance." It closely resembles the crocodile, which is threatened; therefore, the alligator remains on the threatened list because of its resemblance to the crocodile and not because its numbers are low.

Most people fear alligators, but they are reported to be shy about human contact. They will bite if threatened and have been known to eat dogs. Young alligators face numerous hazards, and their mortality rate is high.

Birds

GREAT BLUE HERON

Great blue herons are common sights across all of Arkansas on the edges of rivers, lakes, streams, swamps, and ditches where they routinely put human anglers to shame in fish-catching abilities. Stilt-legged and long-necked, they appear prehistoric and have a loud, guttural squawk. The largest of North American herons, the great blue stands four feet tall and sports two stylish black plumes on its primarily white head and has a long spear-shaped bill. Its neck is brownish-gray, and it has two black shoulder patches. The underside is predominantly dark, and the remainder of the plumage is grayish blue. Immature birds are duller with a dark cap and no plumes.

Great blue heron rookery on the White River near Guion, 1993

Keith Sutton.

Great blues are wary wading birds that spend most of their time motionless in the shallows, waiting for fish to get close enough for them to strike. They bend their long neck into a flattened S-curve while stalking fish or while flying. When they see a fish, the heron stabs its long bill quickly into the water and snaps it shut on the prey. Fish make up the bulk of their diet, although they also eat aquatic invertebrates, small aquatic animals such as baby alligators, and small mammals such as mice.

Great blues are common, permanent residents throughout Arkansas, and their numbers are supplemented by migrating birds from the north and south. Their numbers appear stable in the uplands, but there is some concern about their population in the lowlands where bottomland hardwood breeding habitat is being destroyed.

CATTLE EGRET

Another bird is not so welcome, at least not by Van Buren area residents. Perhaps the cattle egret would be looked on more kindly if six thousand of them did not descend on Van Buren and set up housekeeping every spring.

The egrets winter in South America but seem to prefer life at Van Buren the rest of the year. Because they are protected federally as a migratory species, the egrets cannot be harmed, under penalty of law. So Van Buren building inspector Joe Wilkins and other city workers are forced to use loud noises, gunshots, and cannons firing blanks to scare them away. Wilkins said the egrets once lived along the Interstate 40 access road, but they finally were frightened away from that location. "But they just came back, and last year they moved into the residential area just off the Azure Hills–Blueberry Hill–North Hills additions" where their droppings caused property damage and created a potential health problem.

Cattle egret rookeries are common throughout the state as are complaints about these birds.

TRUMPETER SWAN

About five hundred trumpeter swans are reported to live in the wild in central North America, and they migrate to slightly milder climates when northern waters freeze, occasionally making it to Arkansas. A few trumpeters began winter visits in 1990 to a small lake east of Heber Springs on property owned by Perry Linder Sr. One of the swans was found dead in December 1993 and another was missing. A two-thousand-dollar reward that quickly rose to twenty-eight hundred dollars was offered in an effort to find the swan killer.

Trumpeters are the largest and scarcest waterfowl in North America. They are one of three species of swans living wild in the United States, according to Game and Fish Commission endangered-species biologist Karen Yaich. All three are similar in appearance from a distance but can be distinguished. All are white birds when mature; juveniles have gray bodies.

The mute swan lives on the East and West Coasts and is not seen in Arkansas. The tundra or whistling swan is more common than the trumpeter and is smaller. Tundra and trumpeter swans sometimes mix during winter migration. The tundra has a 72- to 80-inch winspan, is about 52 inches long and weighs from 13 to 18 pounds. The trumpeter, on the other hand, has an 84- to 96-inch wingspan, is about 60 inches or longer and weighs 21 to 30 pounds.

Yaich said the difference in voices may be the best way to identify a swan. The trumpeter has a resonant, sonorous bugle-like call, loud and low-pitched, while the tundra has a high-pitched, quavering "oo-oo-oo" call, accented in the middle. Trumpeters also frequently bow their heads, a form of communication, while calling. Tundra swans do not do this.

INTERIOR LEAST TERN

This endangered species is a summer resident of Arkansas. Because it nests on low sandbars, sometimes whole colonies of least terns are destroyed by high water.

Interior least tern chicks

These chicks were found on a sandbar during a 1994 survey on the Arkansas River near Russellville.

Keith Sutton.

For example, a sandbar near Fort Smith that had a rookery in June 1981 was underwater the next year.

A survey of the Arkansas and White Rivers in June and July 1985 found eighty adult birds, eighty-six juveniles and forty eggs at five separate rookeries on the Arkansas River but none on the White. Other nesting populations are found on large sandbars on the Mississippi River.

BALD EAGLE

Largely through the efforts of volunteer Jane Gulley of Little Rock and the commission, the plight of the endangered bald eagle permeated the consciousness of Arkansans to an unprecedented degree in the 1980s. This does not mean, of course, that eagles are safe from man, and several are killed or injured annually in the state despite the fact they are protected under three federal laws. In fact, Enforcement Division chief David Herman reported in March 1987 that a sixth eagle had been killed in Arkansas, and he complained that the Game and Fish Code did not address eagles specifically. The commission remedied this, adopting a regulation for a state fine of up to one thousand dollars and suspension of any hunting license for two years for killing a raptor. Willful eagle killings were down substantially in Arkansas by the mid-1990s.

The Bald Eagle Act specifies that half of any of the maximum five-thousand-dollar fine collected on a conviction goes to whoever supplies the information leading to the arrest. The commission offers a five-hundred-dollar reward to those furnishing information leading to a conviction for the violation of any major game or fish law, and the National Wildlife Federation has a standing five-hundred-dollar reward for information on eagle killers. The federal Endangered Species Act provides a maximum fifty-thousand-dollar fine and one year imprisonment for those convicted under it. Eagles also are protected under the Migratory Bird Treaty.

In just a few years, the bald eagle became a familiar winter sight in Arkansas, and the state ranks among the top ten in the country in the number of sightings. It is generally known that eagles congregate around big lakes and along the Arkansas River. In fact, there is an annual eagle count sponsored by the National Wildlife Federation and the Raptor Research Foundation along a twenty-four-mile stretch of the Arkansas River from Dardanelle Lock and Dam near Russellville to Lock and Dam No. 9 near Morrilton.

Severe weather can make eagle watching easier. For example, in January 1984, ice covered much of the state, and bald eagles took advantage of the open water and congregated along the Arkansas River. Eagle survey participants aboard the Corps of Engineers's boat *Arkoma* recorded 142 eagles between Lake Dardanelle and Morrilton—a far cry from the 42 spotted in 1983. In 1985, the number of eagles sighted on the river soared to 685—nearly double the 368 recorded in 1979. And by 1986, the survey recorded 953 bald eagles seen in Arkansas. There was an unexplained surge of 1,606 bald eagle sightings in Arkansas in 1992, 335 of them along the Arkansas River. This was far above the average of 959 for the previous six years.

Enforcement chief with dead eagle

David Herman, former chief of the Arkansas Game and Fish Commission's Enforcement Division, with a bald eagle killed by poachers in March 1987.

Arkansas Game and Fish Commission.

Another eagle hang-out was discovered—one not along a major body of water. Doug James, a zoology professor at the University of Arkansas, stood almost in one spot during the 1985–86 winter and counted 115 eagles in one day. He was in extreme western Benton County, south of Maysville and slightly to east of the Oklahoma line (Mosby 1986). Maurice Lous, a local farmer, said the first eagle came in 1979. In 1980, Lous noted four eagles, then 19 in 1981, 33 in 1982, 51 in 1983, and 74 in 1984. The birds feed on chicken carcasses tossed from poultry houses in the area. An alternative food supply is the after-birth of calving cows.

Forty raptors died of botulism in late 1992 in northwest Arkansas after feasting on contaminated, illegally dumped poultry carcasses. Heavy, well-publicized state fines and U.S. Fish and Wildlife Service involvement produced a cleanup in poultry carcass disposal, including revised rules phasing out the use of pits after Springdale-based Tyson Foods, the world's largest poultry processor, began operating a plant near Scranton that produced animal feed from dead birds.

In late 1995 and early 1996 when the carcasses of 29 bald eagles were found around DeGray Lake—the largest eagle die off in the nation's history. The cause was natural but could never be identified despite efforts of state and federal scientists. Tissue from some of the carcasses was frozen in hopes that future scientific knowledge would pinpoint the cause. DeGray continued to have about 15 eagles despite the deaths, and there were 106 of the raptors at Lake Greeson and 75 at Lake Ouachita. A similar die off occurred again in 1996–97.

The first bald eagle nest in Arkansas since 1930 to produce young was reported

Young bald eagles in nest

These seven-week-old Arkansas bald eagles were banded as part of a 1980s research project to determine where the young birds go after leaving the nest. The characteristic white plumage does not develop until the birds are four to five years old.

Arkansas Game and Fish Commission.

in 1982, and seven eagle nests (only one productive) found in Arkansas in 1985 were considered an encouraging sign that their population was doing better. Eighteen nests were considered to be active in Arkansas in 1993, and two more were being built.

The commission started a Bald Eagle Hacking Program on the Buffalo National River in 1982. Young eagles from Minnesota and Wisconsin were brought to the state where they were raised in a tall hacking cage facility. A hacking cage is a tall structure with a cage on top in which young birds of prey are raised to adulthood for release in the wild. The young birds are fed by means of a trays on pulleys to minimize their contacts with humans. Biologists hoped released eagles would return to raise their young in Arkansas and eventually reestablish breeding populations.

About two hundred golden eagles are reported in Arkansas annually. Biologist believe this to be low because state volunteers center their activities along bodies of water where bald eagles are more common. Endangered species specialist Sam Barkley likened scanning Arkansas's woodlands for golden eagles to "finding a diamond at Crater of Diamonds State Park." In fact, the latter may be more productive.

FALCONS

Falcons include merlins, kestrels, and peregrines. Merlins, or pigeon hawks, are seen rarely in Arkansas, with sightings usually occurring near rice fields, reservoirs, or other open expanses during migration or winter.

American kestrels, also called sparrow hawks, reside throughout Arkansas year-round but are most numerous in fall and winter when resident birds are joined by migrants. Nesting occurs from early March through June, usually in tree cavities or outbuildings.

The Arctic peregrine falcon, which occurs in Arkansas, is listed as endangered by the U.S. Fish and Wildlife Service and by Arkansas. Formerly called duck hawks,

Peregrine falcon

In 1993, five fledgling peregrine falcons from Minnesota were given a new home at the Arkansas Power and Light Company power station on the White River in Independence County. They were acclimated to their new area in a hacking station three hundred feet above the ground, then released when ready to fly. Three survived and were often seen flying near the White and Black Rivers.

Mark Wilson.

peregrines are rare migrants and winter visitors in Arkansas. Peregrines have declined throughout their range, in large part because of chemical pollution of their environment. The only published report of a nest in Arkansas for the Arctic peregrine falcon before the 1990s seemed to be one on a bluff along the Little Red River in 1888. The site was inundated by Greers Ferry Lake.

Entergy Corporation, formerly Arkansas Power and Light Company, joined with the Game and Fish Commission in 1994 to bring six baby peregrine falcons from Minnesota to Little Rock (with help from other companies and individuals). The birds were released from their cage high on Little Rock's TCBY Tower on July 16, 1994, and biologists began cutting back on the food they were getting from humans to encourage them to follow their hunting instincts and fend for themselves. Bob West, a biologist with Entergy, reported:

> They've been chasing each other quite a bit. They sometimes fly toward one another and touch feet. This helps them to learn to grab prey in mid-air, which they'll soon be doing with pigeons. I couldn't be happier. They're doing splendid.

Workers in downtown Little Rock who have remained aware of the peregrines delight in looking for the birds riding the air currents above the skyline and perching atop the TCBY Tower, the various bank buildings, and the Tower Building. Early morning and late afternoons are the best time to spot them.

HAWKS

As Apple indicated in his works, relations between man and birds of prey have not always been amicable, though the Native American had a view closer to the

Dead hawks on fence

Wholesale slaughter of hawks, eagles, and other birds of prey was common in Arkansas for many years. It was even encouraged by the Game and Fish Commission. It wasn't until the mid-1960s and early 1970s—a period of heightened environmental consciousness—that these birds began receiving the help and protection they so richly deserve.

Arkansas Game and Fish Commission.

one held today. Plains Indians saw hawks as symbols of fierceness and power, and the Quapaw Indians of Arkansas considered them descendants of the gods.

Hawks found in Arkansas are

Rough-legged hawks, though extremely uncommon in Arkansas, sometimes winter in the state during the coldest weather.

Swainson's hawks are from the western plains but occasionally may be seen in pairs in western Arkansas, especially during spring and fall migrations.

Red-tailed hawks, often called chicken hawks, are the most common and best known in Arkansas as they perch on trees, telephone poles, or fence posts and soar over deciduous forests and open country.

Sharp-shinned hawks, which are difficult to distinguish from Cooper's hawks, are uncommon migrants and winter residents from September to May.

Cooper's hawks are seen in Arkansas year-round.

Broad-winged hawks are known for their migrations with single flocks of up to several thousand soaring in Arkansas skies in April and again in September and October.

Red-shouldered hawks, once quite common in Arkansas, have declined because of wetland clearing and drainage and occur in small numbers statewide during all seasons and nest from late February through June.

Northern harriers, also known as marsh hawks, which arrive in the state from their northern breeding grounds in mid- to late-August and stay until March or April, and occur statewide as one of the most common hawks in the Arkansas River Valley and in eastern farmlands.

Young osprey on a nest

Ospreys were almost wiped out by DDT, which caused eggshell thinning. But following the ban on DDT in the 1970s, the birds recovered. Ospreys began nesting in Arkansas in 1983 after an absence of forty years.

Keith Sutton.

OSPREY

In its magazine, the Game and Fish Commission gave a warm welcome to a bird that returned to Arkansas in 1983 after an absence of forty years. It is the osprey, often called the fish hawk because it feeds primarily on fish. The cause of the osprey's decline was linked to the use of insecticides such as DDT, which had accumulated in the fatty tissue of the fish the birds ate. Because the insecticide interferes with calcium production in bird eggshells, ospreys could not raise any young for years and eventually died out. This began to change in the 1970s with the banning of DDT. A family of ospreys began nesting in 1983 in the Big Lake National Wildlife Refuge, but apparently no eggs hatched. Another nest at nearby Mallard Lake, also in Mississippi County, produced young birds in 1984. The birds also have been seen in Arkansas from Beaver Lake in the northwest to Union County in the south. Their breeding grounds range from the north (Alaska, Yukon, and Northwest Territories) and along the coast as far south as Mexico.

Humans often mistake ospreys for eagles. Ospreys are the only raptor that routinely dives into the water to catch fish. Other birds of prey grab fish from near the surface or feed on floating dead fish. Also, ospreys are not quite as large as eagles. Their wingspans range from four and one-half to six feet, and their body length is twenty-one to twenty-four and one-half inches. On top, they are predominantly dark brown—nearly black—with a purplish hue that sometimes is apparent in strong sunlight. On the underside, white dominates, especially during flight. Their head is white and resembles that of a mature bald eagle; however, ospreys have a conspicuous black band through the eye and across the cheek. Male and female ospreys look alike, although females often are slightly larger.

An osprey nest resembles that of an eagle. It is a large mass of sticks, sometimes with sod included. Most osprey nests are in dead trees ten to sixty feet above ground, but some have been found on rocky outcrops and on the ground. Two to four eggs are laid, most commonly three. The eggs are a colorful pinkish white spotted heavily with reddish brown and dark brown dots. Both sexes incubate the eggs, but generally the female stays on the nest most of the time and eats fish brought to her by the male. Furnishing food is a demanding job, particularly after the clutch hatches because three young eat an average of six pounds of fish a day. Ospreys are fierce in defending the nest and may attack any intruder who threatens it.

No specific osprey management is done in Arkansas, but the bird is protected, and its numbers have rebounded well since DDT was banned.

OWLS

Nine of the eighteen species of owls that occur in North America are known to Arkansas, though the snowy owl has not been seen in the state in more than thirty years. Barn, screech, great horned, and barred owls are permanent Arkansas residents. The short-eared owl is an uncommon winter resident, and the burrowing, long-eared and saw-whet owls are seen only rarely in Arkansas, usually during late fall or winter.

The following is a list of owls seen in Arkansas:

Long-eared owl—reported in Arkansas only eleven times since 1962, always between September and May. It is highly secretive, living in dense pine or mixed forests.

Eastern screech owl—a common year-round Arkansas resident, living in open woodlands near creeks, marshes, and fields. Destructive plant-eating insects compose more than 80 percent of its diet, in addition to crayfish, minnows, wasp larvae, spiders, frogs, birds, and rodents.

Great horned owl—the largest of Arkansas's owls. It lives year-round in woodlands and is more likely to be heard than seen because of its soft, calm hoots.

Burrowing owl—a sporadic winter resident in Arkansas's Delta from late fall through spring. It normally inhabitants western grasslands.

Nongame biologist banding juvenile barn owl

Arkansas Game and Fish Commission biologist Karen Rowe bands a juvenile barn owl from a nest in a grainery near Stuttgart. Barn owls were the focus of intensive management efforts during the 1980s and early 1990s.

Keith Sutton.

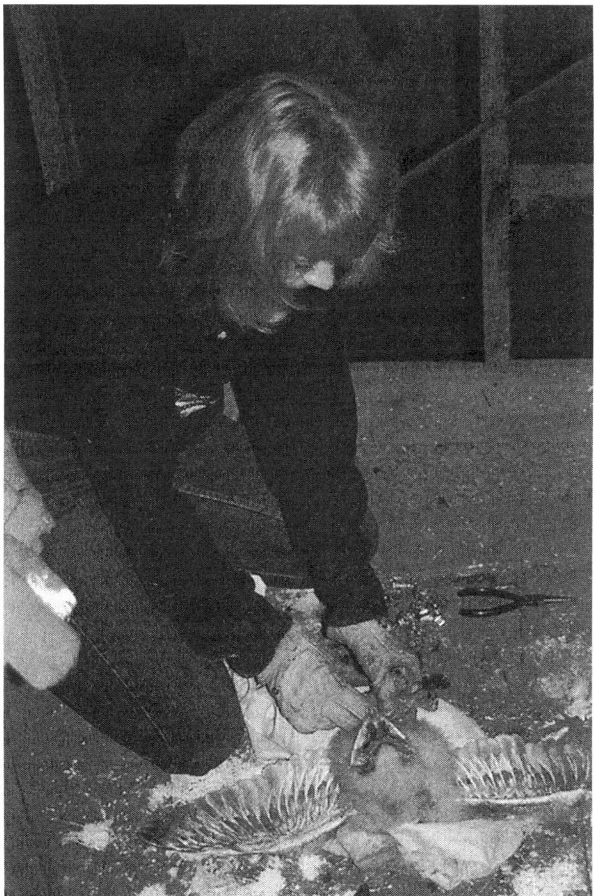

Barred owl—the most vocal of owls and year-round resident of Arkansas bottomland forests.

Short-eared owl—often mistaken for a harrier (marsh hawk). This owl is a migrant and winter resident of Arkansas. They are most common on the Grand Prairie and often are found in association with blackbird roosts, one of their principal items of prey. Active both night and day.

Northern saw-whet owl—this diminutive owl is a rare transient and winter visitor. Its note resembles "the nerve-wracking sound of a saw being filed and may be repeated 100 times."

Common barn owl—a year-round Arkansas resident, welcomed by farmers because they are excellent mousers.

Barn owls are listed as endangered in Missouri and Tennessee and are considered threatened in other neighboring states—but not in Arkansas, perhaps because "We really know very little about the status of the barn owl in Arkansas," Game and Fish Commission biologist Karen Yaich said in 1990. In 1989, the Game and Fish Commission began the first phase of a study to remedy this. Wanted posters went up in the state for barn owl sightings and a network of volunteer investigators was established to check promising reports. Barn owls appeared to be concentrated in the wetter, flatter areas of the state, such as the Delta and the Little River and Red River bottoms in southwest Arkansas. More and more barns are being torn down as agricultural practices change, depriving the owl of one of its favorite haunts. Since the program began, Yaich has maintained nearly three dozen nest boxes in farm outbuildings in various parts of the state. She called on the public to provide more boxes for the owls.

All owls and birds of prey are protected by state and federal laws.

RED-COCKADED WOODPECKER

The red-cockaded woodpecker has received considerable attention in Arkansas in recent years, much of it stemming from a research project by University of Arkansas professor Doug James and graduate student Fred Burnside in 1979. The study, supported by the commission through federal funds, revealed that as of 1979, "89 percent of all active cavity colony sites are on lands controlled by the private timber industry . . . due to current intensified management practices on private holdings the acreage of mature pine forests of the kind required by the red-cockaded woodpecker is steadily declining." Officials of the state's major timber companies have fretted openly about how to avoid problems with the federal government because they own most of the woodpecker's remaining habitat, thus helping to give the bird publicity. The red-cockaded woodpecker was placed on the U.S. Fish and Wildlife Service's endangered species list in 1970. At the time, about 3,000 to 10,000 were estimated to remain in scattered colonies in the southeastern United States, including 200 to 350 in Arkansas (Neal 1981). In the early days, red-cockaded woodpeckers could be found almost anywhere in the state where there were mature pine stands. As Arkansas's old-growth pine timber disappeared, so has this woodpecker.

The red-cockaded woodpecker's name is deceptive. Even wildlife field workers

Red-cockaded woodpecker

Arkansas Game and Fish Commission.

tend to confuse the bird with others that are still abundant, such as the red-headed, red-bellied, and downy woodpeckers. Males of all these birds have red on their heads, but unless a red-cockaded is in the hand, the male's tiny red cockades are virtually invisible. The best field mark is the white cheek patch and chattery voice.

What makes the red-cockaded woodpecker unique—and also has hastened its demise—is its specific habitat requirements. Whereas most woodpeckers excavate their roost/nesting cavities in a dead tree, the red-cockaded chooses a living pine infected with a fungus called red heart, a disease that does not strike until a tree is old and overmature. As the infected tree reaches an advanced age, the woodpecker starts to excavate a cavity, a process consuming several months to more than a year.

The bird is not prolific, either. They live in colonies of two to nine birds, but only one pair breeds each year. The others serve as helpers, excavating new cavities when necessary and assisting the parents in feeding the young.

In a 1981 study, 413 active red-cockaded woodpecker cavity trees were located in Arkansas. Of these, 85 percent were on lands owned by private industry or private individuals. Felsenthal National Wildlife Refuge had 10.4 percent of the active cavities, and 4.6 percent were in the Ouachita National Forest. The Forest Service is managing old-growth timber areas for the red-cockaded woodpecker under the revised 1990 long-range management plan for the Ouachita National Forest.

Game and Fish Commission biologist Rex Roberg reported in 1995 that Arkansas's red-cockaded woodpecker population was at least stable because the Ouachita National Forest and timber companies, particularly Potlatch Corporation, were doing a "great job" of protecting their habitat.

Male bluebird

Leonard Lee Rue III.

EASTERN BLUEBIRD

The bluebird is another species that is in trouble—not endangered or threatened—but troubled. Its numbers have declined drastically in the last fifty years—as much as 90 percent, according to some studies (Mosby 1985).

In an effort to reverse this trend, the Game and Fish Commission has a cooperative program involving many of the state's vocational-technical schools. Woodworking students build bluebird nesting boxes that commission employees and volunteers put up around the state.

The bluebird's principal problem is a familiar one by now—loss of habitat. Karen Yaich, Game and Fish Commission biologist, explained "bluebirds traditionally used decayed trees and rotted wooden fence posts for nesting cavities. Now metal fence posts have replaced wooden ones." The use of pesticides and the increase in English sparrows and starlings also have hurt bluebirds.

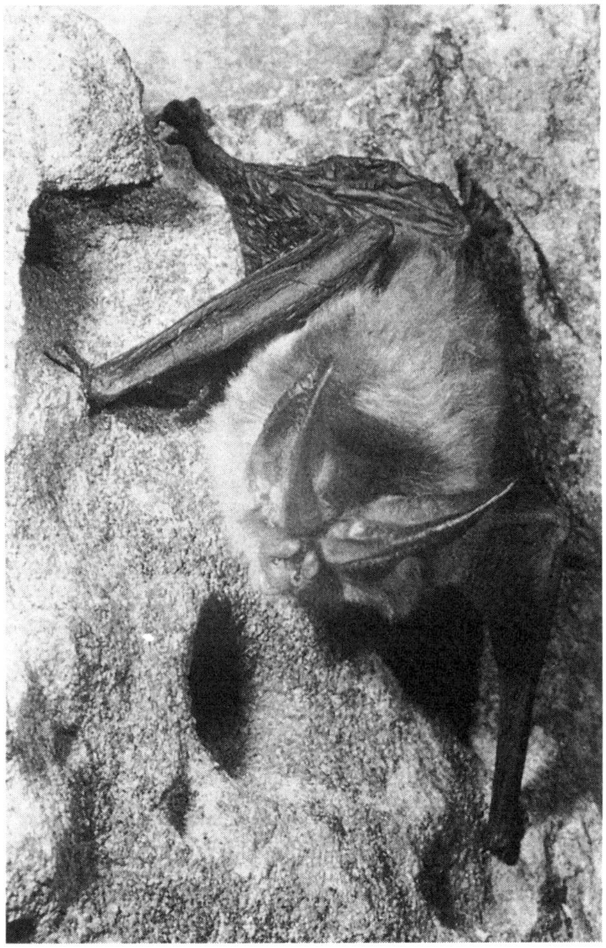

Big-eared bat

The Ozark big-eared bat is one of three endangered Arkansas bat species. Only seventeen hundred remain. Approximately fourteen hundred inhabit a few caves in eastern Oklahoma. The rest live in two Arkansas caves.

Arkansas Game and Fish Commission.

Mammals

BATS

Bats have been around for 60 million years and are found on every continent except Antarctica. With more than one thousand species worldwide, they are second among mammals in number of species only to rodents. Forty bat species are native to the United States, and sixteen occur in Arkansas. Those in Arkansas can be divided broadly into two categories—tree bats and cave bats.

The tree bats are

> Evening bat—the common city dwelling bat in Arkansas. Small and brown, it inhabits buildings and tree cavities during summer. Its winter range is virtually unknown.
>
> Hoary bat—Arkansas's largest. It often migrates in large groups.
>
> Red bat—one of the state's most common bats. It hangs on tree leaves during daytime but is seldom seen. Perhaps more than any other species, this bat feeds on insects attracted to lights.
>
> Seminole bat—until recently found only in a few locations along the Coastal Plain. Now it's moved into the Ouachita Mountains. Little is known about it in Arkansas.
>
> Silver-haired bat—weighing less than a half ounce. This bat is found throughout most of the state. It is a woodland species but occasionally clings to the sides of homes or outbuildings.

There are 11 species of cave bats in Arkansas. Three—Rafinesque's big-eared bat, the Southeastern bat, and the Brazilian freetailed bat—are cave-dwellers in other parts of their range but are seldom found in Arkansas caves. The other cave-dwellers are

> Little brown bat—common in the Arkansas Ozarks and seldom reported in other areas of the state.
>
> Eastern pipistrelle—small bats with tri-colored, orangish fur. "Pips" are common in bat swarms at cave entrances in late summer.
>
> Small-footed bat—rarely encountered. Fewer than twenty have been sighted in caves in northern Arkansas.
>
> Keen's bat—well distributed throughout Arkansas's interior highlands, Keen's bat is distinguished from other myotine species by its longer ears. Hibernates in mines and caves Summer roosts range from buildings to trees.
>
> Big brown bat—one of Arkansas's most common "house" bats. Favored summer roosts are occupied buildings, where hundreds of females form maternity colonies in attics, walls, and other inaccessible places.

Three of the cave-dwelling bats are endangered species. They are

> Gray bat—95 percent of the 1,658,000 known gray bats hibernate in just eight caves—three each in Tennessee and Missouri and one each in Kentucky and Arkansas. The Arkansas cave houses about 250,000 bats, representing more than 15 percent of the total population. The commission and several cooperating state and federal agencies moved to protect caves important to this species by fencing off and gating cave entrances and putting up signs to reduce human disturbance in the caves. The overall population is said to be stable and perhaps even increasing.
>
> Ozark big-eared bat—intolerant of human disturbance, populations of this bat at major Arkansas hibernating caves and maternity caves are believed to be declining. In recent years, the Arkansas population has been as low as twenty, making it the state's most imperiled bat. About 80 percent of the known population occurs in Arkansas. The Nature Conservancy purchased the maternity cave and turned it over to the National Heritage Commission for protection; the Conservancy has a cooperative landowner agreement at the cave.
>
> Indiana bat—The largest number known hibernating in any single cave in Arkansas is about 1,850. Only 3,022 are believed to hibernate in the state, based on a 1982–83 survey. This is a 47 percent decline over just two years. In caves, these bats hibernate rather densely—about 260 bats per square foot. Other than in Ozark caves, the bats hibernate in six locations in Missouri, Indiana, and Kentucky. There are no known maternity sites in Arkansas.

NINE-BANDED ARMADILLO

As anyone who regularly drives Arkansas's highways knows, the state has been invaded by the armadillo.

Actually, it has been a slow migration for this native of Central and South America. No one is certain how long the nine-banded armadillo has lived in Arkansas. The first recorded armadillo report dates to 1921, and it now lives in every county, though there are fewer of them in the chillier northern regions of the state. In the late 1950s, two Little Rock boys caught an armadillo on a city street, and some of the mammals had been seen as far east as El Dorado.

Armadillo

Leonard Lee Rue III.

The armadillo's diet consists primarily of earthworms, grubs, ants, grasshoppers, centipedes, beetles, and a variety of ground-dwelling invertebrates. It is believed widely that armadillos destroy quail eggs and other ground-nesting birds, but experiments with caged animals and small eggs seem to disprove this assumption.

The little armor-covered animal is considered to be harmless to man and to other beasts. In fact, scientists recently discovered that it may harbor the secret that will lead to a cure of leprosy in humans.

FERAL HOGS

There are no true wild hogs in Arkansas. The Arkansas wild hog is a feral pig—a descendant of domestic hogs that escaped or were turned out to range freely. Before the advent of fencing, feral hogs were found statewide. The only sizable populations now are in southern and western Arkansas, particularly the Caney Creek Wildlife Management Area near Mena, and the Felsenthal and Overflow National Wildlife Refuges near Crossett.

Because feral hogs are not game animals, there is no season for hog hunting. However, the U.S. Fish and Wildlife Service, which governs the Felsenthal and Overflow refuges, considers the feral hog a nonnative species that should be kept at manageable levels to minimize conflicts with native wildlife, and there is a liberal hog hunting policy in effect on the two refuges.

At the insistence of the late Commissioner Maurice Lewis of Magnolia, the commission asked the 1993 Arkansas General Assembly to repeal or modify what it considers "outdated laws" concerning feral hogs or give it the authority to regulate them. The commission said in its resolution,

> There is a growing concern among land, wildlife, and forest managers, the state's livestock industry, and the scientific community concerning the potential negative impacts of wild hog populations on human and livestock disease transmission, agricultural crops, wildlife and wildlife food resources, reforestation, and natural plant communities, and free roaming hogs have caused and continue to cause depredation problems on certain forest tree seedlings, agricultural crops, rangelands, lawns, golf courses, and disruption to natural plant communities through rooting, wallowing and trailing.

Arkansas hunter with a "razorback" hog

Arkansas Game and Fish Commission.

Appendixes

APPENDIX I

Arkansas Game and Fish Commission Members

Hal Freeman of Cotter; qualified July 14, 1945, for term expiring June 30, 1946.

Dr. Dan Hamm of Clarksville; qualified July 14, 1945, for term expiring June 30, 1947.

Judge G. E. Keck of Blytheville; qualified July 14, 1945, for term expiring June 30, 1948.

Roger Crowe of Stuttgart; qualified July 14, 1945, for term expiring June 30, 1949.

Roy Perryman of Calico Rock; qualified July 14, 1945, for term expiring June 30, 1950.

Harry G. Barr of Fort Smith; qualified July 14, 1945, for term expiring June 30, 1951.

H. T. Patton of Camden; qualified July 14, 1945, for term expiring June 30, 1952.

Dr. C. W. Dellinger; ex-officio as chairman of the Department of Zoology, University of Arkansas.

Hugh A. Hackler of Mountain Home; qualified August 22, 1946, for term expiring July 1, 1953. Succeeded Freeman.

Armil Taylor of Clarksville; qualified July 3, 1947, for term expiring July 1, 1954. Succeeded Hamm.

***John Campbell** of Oneida; qualified July 8, 1949, for term expiring July 1, 1955. Succeeded Keck.

F. H. McCormack of Stuttgart; qualified July 1, 1949, for term expiring July 1, 1956. Succeeded Crowe.

Charles C. Snapp of Walnut Ridge; qualified August 24, 1950, for term expiring July 1, 1957. Succeeded Perryman.

Dr. Joe Shuffield of Little Rock; qualified June 28, 1951, for term expiring July 1, 1958. Succeeded Barr.

Ed Watkins of Mena; qualified August 21, 1952, for term expiring July 1, 1959. Succeeded Patton.

Dr. J. H. Burge of Lake Village; qualified July 25, 1953, for term expiring July 1, 1960. Succeeded Hackler.

Earl Wells Sr. of Helena; qualified April 9, 1954, for term expiring July 1, 1955. Succeeded Campbell.

Joe Robinson of Springdale; qualified July 2, 1954, for term expiring July 1, 1961. Succeeded Taylor.

Charlie Adams of Hughes; qualified July 1, 1955, for term expiring July 1, 1962. Succeeded Wells.

Riley N. Donoho of Fort Smith; qualified July 9, 1956, for term expiring July 1, 1963. Succeeded McCormack.

***Truman Baker** of Searcy; qualified July 1, 1957, for term expiring July 1, 1964. Succeeded Snapp.

***Ben M. Hogan Sr.** of Little Rock; qualified August 19, 1958, for term expiring July 1, 1965. Succeeded Shuffield.

Gus Pugh of Portland; qualified July 6, 1959, for term expiring July 1, 1966. Succeeded Watkins.

*__Wayne Hampton__ of Stuttgart; qualified August 18, 1960, for term expiring July 7, 1967. Succeeded Burge.

J. Sam Holt of Fulton; qualified January 17, 1961, for term expiring July 1, 1964. Succeeded Baker.

★**Verl Hudspeth** of Harrison; qualified July 10, 1961, for term expiring July 1, 1968. Succeeded Robinson.

Stanley McNulty of Pine Bluff; qualified May 14, 1962, for term expiring July 1, 1967. Succeeded Hampton.

Ernest Hogue of Weiner; qualified July 30, 1962, for term expiring July 1, 1969. Succeeded Adams.

Newt L. Hailey of Rogers; qualified July 1, 1963, for term expiring July 1, 1970. Succeeded Donoho.

Ben M. Hogan Jr. of Little Rock; qualified January 15, 1964, for term expiring July 1, 1965. Succeeded Hogan Sr.

Raymond R. Farris of Biscoe; qualified June 13, 1965, for term expiring July 1, 1971. Succeeded Holt.

Edward Gordon of Morrilton; qualified July 8, 1965, for term expiring July 1, 1972. Succeeded Hogan Jr.

Thomas Robert Pugh of Portland; qualified June 28, 1966, for term expiring July 1, 1973. Succeeded Gus Pugh.

Lloyd McCollum of Stuttgart; qualified July 5, 1967, for term expiring July 1, 1974. Succeeded McNulty.

Neal "Sox" Johnson of Nashville; qualified October 30, 1967, for term expiring July 1, 1968. Succeeded Hudspeth.

Joe D. Scott of Nashville; qualified July 12, 1968, for term expiring July 1, 1975. Succeeded Johnson.

Dr. P. M. Johnston, serving ex-offico as chairman of the Department of Zoology, University of Arkansas. Succeeded Dr. Dellinger.

Ralph B. Griffin of Jonesboro; qualified July 11, 1969, for term expiring July 1, 1976. Succeeded Hogue.

Homer Circle of Rogers; qualified March 5, 1970, for term expiring July 1, 1971. Succeeded Farris.

R. A. Nelson of Blytheville; qualified July 6, 1970, for term expiring July 1, 1977. Succeeded Hailey.

Guy Fenter of Charleston; qualified September 10, 1971, for term expiring July 1, 1978. Succeeded Circle.

Dr. Ralph Bowers of Harrison; qualified December 13, 1972, for term expiring July 1, 1979. Succeeded Gordon.

Michael F. Mahony of El Dorado; qualified August 7, 1973, for term expiring July 1, 1980. Succeeded Pugh.

W. Fay Wright of North Little Rock; qualified July 24, 1974, for term expiring July 1, 1981. Succeeded McCollum.

★**Henry Moore III** of Texarkana; qualified July 3, 1975, for term expiring July 1, 1982. Succeeded Scott.

★**Kaneaster Hodges** of Newport; qualified July 16, 1976, for term July 1, 1983. Succeeded Griffin.

Hays L. Sullivan of Burdette; qualified July 15, 1977, for term expiring July 1, 1984. Succeeded Nelson.

Alton R. "Rick" Hampton of Stuttgart; qualified January 9, 1978, for term expiring July 1, 1983. Succeeded Hodges.

Dr. George Cole Jr. of Fayetteville; qualified September 11, 1978, for term expiring July 1, 1985. Succeeded Fenter.

Hilary Jones of Pruitt; qualified July 27, 1979, for term expiring July 1, 1986. Succeeded Bowers.

Dr. Duncan W. Martin, serving ex-officio as chairman of the Department of Zoology, University of Arkansas. Succeeded Dr. Johnston.

N. Casey Jones of Pine Bluff; qualified July 1, 1980, for term expiring July 1, 1987. Succeeded Mahony.

Ross Whipple of Arkadelphia; qualified March 24, 1981, for term expiring July 1, 1982. Succeeded Moore.

Beryl Anthony Sr. of El Dorado; qualified July 1, 1981, for term expiring July 1, 1988. Succeeded Wright.

Frank Lyon Jr. of Little Rock; qualified July 14, 1982, for term expiring July 1, 1989. Succeeded Whipple.

Tommy Sproles of Little Rock; qualified July 26, 1983, for term expiring July , 1990. Succeeded Hampton.

William Brewer of Paragould; qualified August 20, 1984, for term expiring July 1, 1991. Succeeded Sullivan.

Perry Mikles of Booneville; qualified August 15, 1985, for term expiring July 1, 1992. Succeeded Cole.

Dr. Charles J. Amlaner Jr., serving ex-officio as chairman of the Department of Zoology, University of Arkansas. Succeeded Dr. Martin.

Mike Cornwell of Danville; qualified February 16, 1987, for term expiring July 1, 1993. Succeeded Hilary Jones.

Harold Ives of Stuttgart; qualified August 25, 1987, for term expiring July 1, 1994. Succeeded N. Casey Jones.

****Maurice Lewis Jr.** of Magnolia; qualified July 11, 1988, for term expiring July 1, 1996. Succeeded Anthony.

David E. Miller of Melbourne; qualified November 8, 1989, for term expiring July 1, 1996. Succeeded Lyon.

Dr. Arnold Kaplan of the University of Arkansas at Fayetteville, served October 1991 to May 1995. Succeeded Dr. Amlaner.

James E. Moore Jr. of Little Rock; qualified July 1990 for term expiring July 1, 1997. Succeeded Sproles.

***Hal Hunnicut** of Conway; qualified July 8, 1991, for term expiring July 1, 1998. Succeeded Brewer.

J. Kirkwood Dupps of Bentonville/Eureka Springs; qualified July 14, 1992, for term expiring July 1, 1999. Succeeded Mikles.

Wilton Robert "Witt" Stephens Jr. of Little Rock; qualified July 2, 1993, for term expiring July 1, 2000. Succeeded Cornwell.

William Bridgforth of Pine Bluff; qualified July 5, 1994, for term expiring July 1, 2001. Succeeded Ives.

*****Pat Stephens (Johnson)** of Stuttgart; qualified July 11, 1994, for term expiring July 1, 1995. Replaced Maurice Lewis Jr.

Richard W. "Rick" Evans of Calion; qualified December 5, 1994, for term expiring July 1, 1998. Succeeded Hunnicut.

Dr. Dwight Talburt of the University of Arkansas at Fayetteville, began serving June 1995. Succeeded Dr. Kaplan.

Marion T. Mccollum of Stuttgart; qualified July 1, 1995, for term expiring July 1, 2002. Succeeded Pat Stephens (Johnson).

James G. Hinkle of Mountain View; qualified July 1, 1996, for term expiring July 1, 2003. Succeeded David Miller.

Lester Sitzes of Hope; qualified July 1, 1997, for term expiring July 1, 2004. Succeeded James E. Moore Jr.

---

*Term not completed.

**Died unexpectedly in March 1994. The respect he earned as a commissioner is demonstrated by the fact the Game and Fish Commission headquarters complex at 2 Natural Resources Drive, Little Rock, was named in his honor.

***Appointed by Gov. Jim Guy Tucker, this world-champion duck caller was the first woman to serve on the Commission.

Source: Secretary of State's Office, State Capitol, Little Rock, Arkansas.

APPENDIX 2
History Of Arkansas Game and Fish Licenses

Fishing Licenses

TYPE OF LICENSE	YEAR	COST TO BUYER
Artificial Bait Fishing	1915	1.00
	1921	1.10
	1937	1.50
Resident Fishing (Name change 1940)	1958	2.50
	1966	3.50
	1976	5.00
	1978	7.50
	1984	10.50
	1995	11.50*
	1997	10.50
Non-Resident Annual Fishing	1919	5.00
	1966	6.00
	1976	8.00
	1978	10.30
	1981	15.50
	1987	20.00
	1991	25.00
	1992	26.00
	1995	30.00
Non-Resident Three-Day Trip (Discontinued 12/31/83)	1981	7.50
Resident and Non-Resident Three-Day Trip (Discontinued 6/30/91)	1984	5.00
	1987	6.50
Non-Resident Seven-Day Trip	1995	15.00
Non-Resident Fourteen-Day Trip	1919	1.10
	1939	2.00
	1959	2.50
	1966	3.50
	1976	5.00
	1978	7.50
	1981	10.50
	1987	13.50

	1992	15.00
	1995	20.00
Resident Three-Day Trip	1992	6.50
	1995	7.50*
	1997	6.50
Non-Resident Three-Day Trip	1992	8.00
	1995	10.00
Resident (Age 65 and Over) Lifetime Fishing	1965	3.50
	1985	10.50
	1995	11.50*
	1997	10.50
Disabled Lifetime Fishing (Discontinued 1992)	1989	10.50
Trout Permit (Resident and Non-Resident until Separated)	1971	2.00
	1977	3.00
	1987	5.00
Non-resident Trout Permit	1994	7.50
Commercial Fishing Permit (Discontinued 12/31/85)	1977	50.00
Commercial Fishing Permit and Sportfishing License	1986	25.00
	1995	26.00
Resident Commercial Fisherman's Helper Permit	1993	25.00
Junior/Senior Commercial Fishing Permit	1986	14.50
Fish Farmer Permit	1964	10.00
	1979	25.00
Fiddler or Slat Trap	1982	12.50
Hoop Nets	1924	25.00
Replaced 1955	1934	1.00
Hoop Nets, Short Replaced 1977	1955	2.50
Hoop Nets, Long Replaced 1977	1955	5.00
Hoop Nets, Other (Discontinued 6/30/60)	1955	7.50
Hoop Nets (Turtle Net Tag Added 1993)	1977	4.00
Resident Fish Dealer	1938	5.00
	1948	10.00

Minnow Trap (Discontinued 6/30/81)	1950	1.00
Minnow Seine Replaced 1987	1950	5.00
Minnow Seine, Trap, Dip, or Lift	1987	5.00
Non-Resident Minnow Dealer (Discontinued 12/31/88)	1955	200.00
Non-Resident Commercial Fish Dealer	1989	250.00
Put and Take Pay Lake	1969	100.00
	1980	150.00
Seines and Nets, Long (Discontinued 6/30/59)	1934	25.00
Seines and Nets, Short	1937	12.50
Replaced 1987	1977	15.00
Seines, Trammel Nets or Gill Nets	1987	15.00
Trot Line	1924	10.00
	1932	2.50
	1933	10.00
	1934	2.50
	1956	1.00
	1971	2.00
Name Change: Trot or Snag Lines, Set Hooks and Lines (Turtle Box Trap Added 1993)	1980	2.00
Non-Resident Turtle Permit (Discontinued 7/20/87)	1964	200.00
Resident Shell Taker	1938	1.00
(Discontinued 12/31/85)	1971	2.00
	1982	5.00
Resident Shell Taker, Buyer, Seller	1986	10.00
(Discontinued 12/31/94)	1993	25.00
Resident Shell Taker, Seller	1995	100.00
Resident Shell Buyer	1995	1,000.00
Shell Taker Helper	1994	25.00
	1995	100.00
Shell Buyer (Discontinued 12/31/85)	1938	50.00
Shell Buyer's Agent	1996	100.00
Non-Resident Shell Taker (Discontinued 12/31/85)	1965	25.00

Non-Resident Shell Taker, Buyer, Seller	1986	50.00
(Discontinued 12/31/89)	1989	300.00
Non-Resident Shell Buyer	1990	50.00
	1993	250.00
	1995	2,000.00
Alligator Farming Permit	1992	500.00
Bullfrog Permit	1969	25.00

Hunting Licenses

TYPE OF LICENSE	YEAR	COST TO BUYER
Resident Hunting	1915	1.00
	1921	1.10
	1937	1.50
	1958	2.50
	1966	3.50
	1976	5.00
	1978	7.50
	1984	10.50
	1995	11.50★
Resident Sportsman's License	1984	17.75
	1988	25.00
	1989★★	17.75
	1990	25.00
	1995	26.00★
Non-Resident Annual All-Game (Discontinued Fiscal Years 1971-1983)	1923	15.00
	1936	25.00
	1966	30.00
	1983	115.00
	1988	150.00
	1994	185.00
	1996	195.00
Non-Resident Deer Tags (Discontinued 6/30/83)	1971	10.00
	1978	15.00
	1980	20.75
	1983	38.50
Non-Resident Annual Small Game	1937	15.00
	1966	20.00
	1976	25.00
	1980	36.50
	1984	38.00
	1988	50.00
	1994	65.00
	1996	75.00

License	Year	Fee
Non-Resident Five-Day Small Game	1971	7.00
	1976	10.00
	1980	15.50
	1984	22.00
	1988	30.00
	1994	40.00
	1996	50.00
Resident Lifetime Sportsman's License	1966	3.50
	1985	17.75
	1988	25.00
	1989**	17.75
	1990	25.00
	1995	26.00*
	1997	25.00
Disabled Lifetime Hunting (Discontinued 1992)	1990	25.00
Lifetime Sportsman's Permit (For Residents of Any Age)	1988	1,000.00
Three-Day Trip All-Game	1992	75.00
	1994	95.00
Five-Day Trip All-Game	1986	78.00
	1988	100.00
	1994	125.00
Resident Bonus Antlerless Deer Permit (Discontinued 4/13/98)	1996	10.00
Non-Resident Bonus Antlerless Deer Permit (Discontinued 4/13/98)	1996	10.00
WMA Deer Permits, Zone Deer Permits, Special Antlerless Deer Permits, and Turkey Permits (Drawn Randomly from Applications)	1993	10.00
Leased Land Permit (Discontinued 6/30/91)	1989	10.00
Leased Land Permit, Big Timber WMA	1992	10.00
	1992	20.00
Leased Land Permit, Casey Jones WMA	1992	10.00
	1992	20.00
Leased Land Permit, Cherokee WMA	1992	10.00
	1992	20.00

License	Year	Fee
Special Hunting Permit	1990	10.00

This permit was replaced 7/1/91 by six separate permits for archery deer, bear, furbearers-trapping, modern gun deer, muzzleloading deer, and turkey. All were discontinued 2/28/95.

Resident Archery	1977	5.00
(Discontinued 1983)	1983	7.50
Non-Resident Archery	1971	10.00
(Discontinued 1983)	1979	20.00
	1980	20.75
Muzzleloading Permit	1978	10.00
(Discontinued 1983)		
Resident Turkey Tag	1981	5.50
(Discontinued 1983)		
Non-Resident Turkey Tag	1971	10.00
(Discontinued 1983)	1978	15.00
	1980	20.75
	1983	38.50
Waterfowl Stamp	1982	5.50
	1988	7.00
	1989**	5.50
	1990	7.00
Non-Resident Waterfowl Permit	1993	12.00
(Discontinued 6/30/96)		
Falconry, Apprentice	1992	25.00
General	1992	25.00
Master	1992	25.00
Non-Resident Fur Taker	1944	50.00
	1994	100.00
Resident Fur Dealer	1938	1.00
	1940	5.00
	1978	50.00
Non-Resident Fur Dealer	1939	25.00
	1940	50.00
	1977	100.00
	1985	200.00

Resident Live Fox and Coyote Permit 1997 100.00

Non-Resident Live Fox and Coyote Permit 1997 250.00

Combined Licenses

TYPE OF LICENSE	YEAR	COST TO BUYER
Resident Combination Sportsman's License	1988	30.25
	1989	35.50
	1989**	28.25

	1990	35.50
	1995	37.50★
	1997	35.50
Resident (Age 65 and Over) Lifetime Combination Hunting and Fishing License	1966	7.00
	1984	28.25
	1988	35.50
	1989★★	28.25
	1990	35.50
	1995★	37.50
	1997	35.50
Disabled Lifetime Combination License (Discontinued 1992)	1990	35.50

Miscellaneous

Dog Tags	1915	1.00
	1919	1.50
	1980	2.50
Commercial Shooting Resort	1956	50.00
	1979	100.00
	1980	150.00
Commercial Game Breeder Permit	1940	25.00
Special Commercial Quail Breeders Permit	1956	25.00
Commercial Non-Game Breeders Permit	1988	25.00
Guide (Discontinued 1933)	1932	10.00
Guide, Fishing (Discontinued 1949)	1934	5.00
Guide, Hunting (Discontinued 1949)	1934	10.00
Guide License, Hunting and Fishing	1950	10.00
(Discontinued 6/30/91)	1987	20.00
Guide License, Hunting	1992	25.00
Guide License, Fishing	1992	25.00
Guide License, Special	1993	150.00
Guide License, Special (Limited to Residents)	1996	150.00
Guide License, Non-Resident: Hunting	1996	150.00
Commercial Boat Dock	1968	10.00
	1988	20.00
	1993	25.00

Boat House, Joint	1967	10.00
Boat House, Single	1967	5.00
Boat Dock Commercial (User Fee) (Discontinued 1979)	1972	250.00
Boat Dock Launch (Discontinued 1986)	1980	100.00
Commerical Dock, Parking, Launching	1986	150.00
Boat Rental (Discontinued 12/31/89)	1986	2.00
Dock Permit, Joint	1969	10.00
Dock Permit, Single	1969	5.00
Alligator Snapping Turtle Farmer/Dealer Permit	1998	100.00

* For two years only.

** A result of Arkansas Supreme Court decision.

Source: Ray Sebren, Chief, Fiscal Division, Arkansas Game and Fish Commission.

APPENDIX 3

Income And Expenditures: Amendment 35 Arkansas Game and Fish Commission

YEAR	GAME PROTECTION BALANCE	INCOME	EXPENDITURES	SURPLUS (DEFICIT)
1945	219,635	315,123	300,788	14,335
1946	460,965	574,147	332,818	241,329
1947	791,592	803,118	472,490	330.628
1948	815,097	851,058	827,554	23,504
1949	872,277	936,590	879,410	57,180
1950	565,693	973,693	1,280,276	(306,583)
1951	307,006	1,000,960	1,259,648	(258,688)
1952	323,562	1,210,617	1,194,060	16,557
1953	378,196	1,354,662	1,300,028	54,634
1954	253,570	1,358,068	1,482,695	(124,627)
1955	639,698	1,993,179	1,607,051	386,128
1956	751,194	1,991,875	1,880,380	111,495
1957	529,594	1,691,968	1,913,567	(221,599)
1958	395,194	2,575,384	2,710,159	(134,775)
1959	418,930	2,607,022	2,583,286	23,736
1960	196,781	2,617,957	2,828,127	(210,170)
1961	269,871	2,398,593	2,392,934	5,659
1962	559,872	2,873,925	2,543,544	330,381
1963	615,043	2,732,629	2,673,067	59,562
1964	331,144	2,690,909	2,959,657	(268,748)
1965	415,039	2,995,381	2,904,405	90,976
1966	1,481,463	3,666,032	2,599,641	1,066,391
1967	2,379,171	3,850,493	2,984,888	865,605
1968	3,152,745	4,067,759	3,294,185	773,574
1969	3,658,669	4,277,823	3,771,900	505,923
1970	3,410,207	4,910,454	5,173,145	(262,691)
1971	2,568,981	5,424,204	6,264,952	(840,748)
1972	2,532,857	5,661,049	5,713,032	(51,983)
1973	2,132,844	5,908,326	6,299,603	(391,276)
1974	1,560,234	5,973,000	6,542,364	(569,364)
1975	1,074,181	6,216,672	6,788,046	(571,374)
1976	1,460,648	7,818,175	7,443,631	374,544
1977	1,279,179	8,115,900	8,298,165	(182,265)
1978	1,818,749	10,390,536	10,251,731	138,805
1979	1,758,109	11,542,929	10,914,972	627,957

1980	2,790,288	12,461,401	11,140,992	1,320,408
1981	3,996,558	13,141,214	12,104,790	1,036,424
1982	1,184,572	13,097,689	12,658,025	439,664
1983	7,145,437	14,828,282	12,771,816	2,056,466
1984	8,005,704	16,292,558	15,215,880	1,076,678
1985	8,429,406	16,221,925	15,191,136	1,030,789
1986	6,819,819	17,192,061	18,436,561	(1,244,500)
1987	3,082,295	17,537,400	19,925,401	(2,388,001)
1988	5,209,408	21,488,275	20,537,417	950,858
1989	3,060,815	22,344,571	24,103,509	(1,758,938)
1990	2,177,590	23,568,535	26,239,550	(2,671,015)
1991	3,944,056	25,320,051	22,881,298	2,438,753
1992	3,404,250	25,599,562	26,365,368	(765,806)
1993	2,597,961	26,762,675	27,073,949	(311,274)
1994	2,470,155	25,259,137	25,331,185	(72,048)
1995	2,794,403	25,984,646	24,734,480	1,250,166
1996	1,962,761	28,785,140	29,778,155	(993,015)

Source: Ray Sebren, Chief, Fiscal Services Division, Arkansas Game and Fish Commission, April 1998.

APPENDIX 4
License Sales: 1985–1997

FISCAL YEAR	QUANTITY	+ OR ()	AMOUNT	+ OR ()
1985	1,227,175		$ 11,725,926.70	
1986	1,300,584	73,409	12,493,110.35	$ 767,191.65
1987	1,316,260	15,676	13,073,105.00	579,994.65
1988	1,284,400	(31,860)	15,405,324.55	2,332,219.55
1989	1,277,882	(6,518)	14,882,860.45	(522,464.10)
1990	1,275,722	(2,160)	14,781,619.95	(101,240.50)
1991	1,323,166	47,444	15,054,112.00	272,492.05
1992	1,325,508	2,342	16,651,846.00	1,597,734.00
1993	1,258,078	(67,430)	16,179,336.00	(472,510.00)
1994	1,265,774	7,896	16,489,754.50	310,418.50
1995	1,272,643	6,669	17,902,378.50	1,412,624.00
1996	1,312,533	39,890	19,307,128.48	1,404,749.98
1997	1,314,737	2,204	19,710,405.50	403,277.02

Source: Fiscal Division, Arkansas Game and Fish Commission, Little Rock, Ark., March 1998.

Bibliography

Books

Allen, Durward L. *Our Wildlife Legacy*. Rev. ed. New York: Funk and Wagnals Company, 1962.

Bass, Sharon M. W. *For the Trees—An Illustrated History of the Ozark–St. Francis National Forests, 1988–1978*. Atlanta, Ga.: United States Department of Agriculture, Forest Service Southern Region, 1981.

Black, John D. *The Management and Conservation of Biological Resources*. Philadelphia: F. A. Davis Company, 1968.

Bloyd, Sunni. *Endangered Species*. San Diego, Ca.: Lucent Books, 1989.

Clepper, Henry. *Professional Forestry in the United States*. Baltimore: Johns Hopkins University, 1971. (Prepared for Resources for the Future.)

Desmarais, Ralph, and Robert Irving. *The Arkansas Grand Prairie*. Stuttgart, Ark.: County Agricultural Museum, 1988.

Donovan, Timothy P. and Willard B. Gatewood. Jr., eds. *The Governors of Arkansas*. Fayetteville: The University of Arkansas Press, 1981.

Dougan, Michael B. *Arkansas Odyssey—The Saga of Arkansas from Prehistoric Times to Present*. Little Rock: Rose Publishing Company, 1994.

Dunlap, Thomas R. *Saving America's Wildlife*. Princeton: Princeton University Press, 1988.

Easterbrook, Gregg. *A Moment on the Earth*. New York: The Penguin Group, Viking Press, 1995.

Gerstaecker, Friedrich. *Western Lands and Western Waters*. London: S. O. Beeton, 1864.

Graham, Frank, Jr. *Man's Dominion—The Story of Conservation in America*. New York: M. Evans and Company, 1971.

Graustein, Jeannette. "Explorations in the Southwest, The Arkansas Journey, 1818–1820." In *Thomas Nuttall, Naturalist—Explorations in America, 1808–1841*. Cambridge: Harvard University, 1967.

Halls, Lowell K., ed. *White-tailed Deer—Ecology and Management*. Harrisburg, Pa.: Stackpole Books, 1984.

Hays, Samuel P. *Conservation and the Gospel of Efficiency—The Progressive Conservation Movement, 1890–1920*. Cambridge: Harvard University Press, 1959.

Holder, T. H. *A Survey of Arkansas Game*. Little Rock: Arkansas Game and Fish Commission, Federal Aid Publication Project 11-R, 1951.

Pope, Judge William F. *Early Days in Arkansas, Being for the Most Part the Recollections of an Old Settler*. Edited by Dunbar H. Pope. Little Rock, Ark.: Frederick W. Allsopp, 1895.

Shabecoff, Philip. *A Fierce Green Fire*. New York: Hill and Wang, 1993.

Stoddard, Charles H. *Essentials of Forestry Practice*. New York: Ronald Press Company, 1968.

Stroud, Hubert B. and Gerald T. Hanson. *Arkansas Geography, The Physical Landscape and the Historical Cultural Setting*. Little Rock, Ark.: Rose Publishing Company, 1981.

Thomas, David Y., ed. *Arkansas and Its People; A History, 1541–1930*. New York: The American Historical Society, 1930.

Trefethen, James B. *Crusade for Wildlife—Highlights in Conservation Progress*. Harrisburg, Pa.: The Stackpole Company, and New York: Boone and Crockett Club, 1961.

Reports and Proposals

Arkansas Elections—1984. Little Rock, Arkansas: Secretary of State; p. 64.

Arkansas Game and Fish Commission, Little Rock, Arkansas:

First Annual Report, March 11, 1915 to June 15, 1916.

Second Annual Report, June 15, 1916 to March 31, 1918.

Third Biennial Report, April 1, 1918 to April 1, 1920.

Fourth Biennial Report, July 1, 1920 to June 30, 1922.

Fifth Biennial Report, June 30, 1922 to July 1, 1924.

Sixth Biennial Report, July 1, 1924 to June 30, 1926.

Seventh Biennial Report, July 1, 1926 to June 30, 1928.

Eighth Biennial Report, July 1, 1928 to June 30, 1930.

Ninth Biennial Report , July 1, 1930 to June 30, 1932.

Tenth Biennial Report , July 1, 1932 to June 30, 1934.

[The] Arkansas Game and Fish Commission. Independent Study Team. Little Rock: Arkansas Game and Fish Commission, January 1970.

Arkansas Natural Area Plan. Little Rock: Arkansas Department of Planning, December 1974.

Arkansas Statewide Study, Phase V. Washington, D.C.: United States Department of Agriculture, Soil Conservation Service, Economic Research Service, Forest Service, in cooperation with the Arkansas Soil and Water Conservation Commission, August 1983.

Characteristics of the Population. Vol. 1, Number of Inhabitants, Chapter A, Arkansas, Part 5, PC80-1-A5. Washington, D.C.: Bureau of the Census, United States Department of Commerce, February 1982.

Expanding the Commitment. Update, North American Waterfowl Management Plan. Washington, D.C.: U.S. Fish and Wildlife Service, Department of the Interior, 1994.

Griffee, Carol. *Environmental Quality Index.* Little Rock: Arkansas Wildlife Federation, 1988.

———. *Environmental Quality Index.* Little Rock: Arkansas Wildlife Federation, 1994.

Highlights in the History of Forest Conservation. Washington, D.C.: United States Department of Agriculture, Forest Service, AIB-83, rev. March 1976.

100 Years of Federal Forestry. Agriculture Information Bulletin No. 402, December 1976. Washington, D.C.: United States Department of Agriculture, Forest Service,

Purvis, George. *A Brief Review of the Duties and Activities of the Game and Fish Commission.* Little Rock: Arkansas Game and Fish Commission, February 1, 1954.

Roberg, Rex. *Summary of Midwinter Bald Eagle Survey in Arkansas, January 1–15, 1992.* Little Rock: Arkansas Game and Fish Commission.

Taylor, Nell. *Report on Income and Disbursements—March 1915 Through June 1957.* Prepared for Judge Lee Miles. Little Rock: Arkansas Game and Fish Commission, August 18, 1957.

Water Resources Development in Arkansas. Little Rock, Arkansas: United States Army Corps of Engineers, Southwestern Division, January 1, 1981.

Minutes

Little Rock, Arkansas: Arkansas Game and Fish Commission.

October 26, 1943; December 20, 1943; February 14, 1944; May 22, 1944; June 19, 1944; September 18, 1944; October 23, 1944; January 15, 1945; February 19, 1945; April 16, 1945; August 20, 1945; January 21, 1946; April 14, 1946; April 15, 1946; May 18, 1946; July 15, 1946; October 21, 1946; February 17, 1947; March 17, 1947; April 21, 1947; July 21, 1947; September 15, 1947; November 17, 1947; December 29, 1947; February 16, 1948; March 15, 1948; April 19, 1948; May 9, 1948; June 21, 1948; September 20, 1948; October 18, 1948; February 21, 1949; July 18, 1949; September 19, 1949; October 17, 1949; December 18, 1949; March 20, 1950; July 17, 1950; September 18, 1950; October 23, 1950; November 20, 1950; April 16, 1951; June 18, 1951; August 1, 1951; August 20, 1951; January 21, 1952; December 15, 1952; February 16, 1953; July 20, 1953; April 18, 1954; July 8, 1956; January 20, 1986; February 17, 1986; April 21, 1986; May 19, 1986; June 16, 1986; July 21, 1986; August 18, 1986; September 19, 1986; November 17, 1986; December 15, 1986; January 19, 1987; February 17, 1987; March 16, 1987; April 13–14, 1987; June 15, 1987; May 18, 1987; July 20, 1987; August 17, 1987; October 19, 1987; December 21, 1987; January 18, 1988; August 22, 1988; November 21, 1988; April 17, 1989; July 17, 1989; August 21, 1989; October 20, 1989; November 20, 1989; December 18, 1989; February 19, 1990; March 26, 1990; April 16, 1990; May 21, 1990; June 25, 1990; August 27, 1990; November 3, 1990; January 3, 1991; February 18, 1991; March 18, 1991; April 15, 1991; May 20, 1991; June 24, 1991; August 19–20, 1991; October 5, 1991; October 14, 1991; November 26, 1991

Articles: Historical Publications

Baker, Russell P. "Jacob Wolf." *Arkansas Historical Quarterly* 36, no. 2 (summer 1979): p. 188.

"Bear Hunting in Phillips County." *Phillips County Historical Quar*terly (June 1977): pp. 32–33.

Browning, Jo Ann. "The History of St. Charles." *Arkansas Historical Quarterly* 12, no. 4 (winter 1953): p. 386.

Buffaloe, W. T. "'Uncle France' Varner." *Arkansas Historical Quarterly* 20, no. 2 (summer 1961): p. 161.

Cathey, Henry, ed. "Extracts from Memoirs of William Franklin Avera." *Arkansas Historical Quarterly* 14, no. 2 (summer 1955): pp. 99–116.

"Chips and Shavings." *Arkansas Historical Quarterly* 11, no. 3 (autumn 1952): pp. 217–18.

Eison, James Reed, ed. "A Letter from Dardanelle to Jonesville, S.C." *Arkansas Historical Quarterly* 28, no. 1 (spring 1969): p. 74.

Evans, Clarence and Liselotte Albrecht, trans. "Friedrich Gerstaecker in Arkansas, "Selections from his *'Striefund Jagdzuege Durch Die Vereingten Staaten Nordamerikas. Arkansas Historical Journal"* 5, no. 1 (spring 1946): pp. 54–56.

Helena World, item, March 16, 1898. Rep. *Phillips County Historical Quarterly*, (June 1977).

Honor, Dr. Albert A. "Bear Hunting." *Phillips County Historical Quarterly* 5, no. 4 (September 1967): pp. 25–26.

Ledbetter, Cal, Jr. "The Office of Governor in Arkansas History." *Arkansas Historical Quarterly* 36, no. 1 (spring 1978): pp. 72–73.

Martel, Glenn. "Origin of Columbia Place Names Reviewed." *Arkansas Historical Quarterly* 11, no. 1 (spring 1952): p. 3.

Martel, Glenn G. "Early Days in Columbia County." *Arkansas Historical Journal* 2, no. 3 (September 1943): pp. 214–16.

McIver, H. M. "Water Bound in Arkansas." *Arkansas Historical Quarterly* 14, no. 2 (summer 1955): p. 152.

Morris, Robert L. "Three Arkansas Travelers." *Arkansas Historical Journal* 14, no. 3 (autumn 1945): p. 226.

Neal, James P. "Wild Animals, Fowls and Reptiles in the Early Days." *Flashback* 8, no. 6 (October 1958): pp. 13–14.

Plaisance, Aloysius. "The Arkansas Factory, 1805–1810." *Arkansas Historical Quarterly* 11, no. 3 (autumn 1952): pp. 184–87.

Poe, Deleta Jacqueline. "Lake Conway, Arkansas' Newest, Finest Playground." *Arkansas Historical Quarterly* 12, no. 2 (summer 1953): pp. 107–11.

Propps, J. J. "My Childhood and Youth in Arkansas." *Arkansas Historical Quarterly* 26, no. 4 (winter 1967): pp. 311–12.

Ross, Margaret Smith. "Herman Davis: Forgotten Hero." *Arkansas Historical Quarterly* 14, no. 1 (spring 1955): p. 51.

Rowland, Mrs. Dunbar. "Life, Letters and Papers of William Dunbar." Jackson, Miss.: Press of Mississippi Historical Society, 1930: n.p.

Simonson, S. E. "The St. Francis Levee and High Water on the Mississippi River." *Arkansas Historical Journal* 6, no. 4 (winter 1947): p. 419.

Smith, Ralph A., trans."Explorations of the Arkansas River by Bernard de La Harpe, 1721–1722," (Excerpts from his journal and instructions.) *Arkansas Historical Journal* 10, no. 4 (winter 1951): pp. 339–63.

Vaulx, Julia R. "Another Early Traveler in Arkansas." *Arkansas Historical Journal* 5, no. 2 (summer 1946): pp. 176–77.

Williams, C. Fred. "The Bear State Image." *Arkansas Historical Quarterly* 39, no. 2 (summer 1980): p. 103.

Williams, W. D., ed. "1820 Letter from Governor Miller Describing Arkansas Territory." *Arkansas Historical Quarterly* 42, no. 4 (winter 1983): p. 351.

Worley, Ted R. "An Early Arkansas Sportsman: D. F. M. Noland." *Arkansas Historical Quarterly* 11, no. 1 (spring 1952): pp. 25–28.

———. "Letters From an Early Arkansas Settler." *Arkansas Historical Quarterly* 11, no. 4 (winter 1952): pp. 327–29.

———. "Story of an Early Settlement in Central Arkansas." *Arkansas Historical Journal* 10, no. 2 (summer 1951): pp. 122–29.

Wynn, Stella. "Howard County Game Refuge and Deer Farm." *Arkansas Historical Quarterly* 2, no. 4 (December 1943): pp. 340–43.

Articles: the *Arkansas Deer,* later the *Arkansas Conservationist*

Allen, W. S. "One of the Old Boys Saw." 1, no. 1 (1924): pp. 9–10.

Beauchamp, D. G. "Fishing Practically All Gone." 3, no. 3 (1926): p. 12.

"Conserving Our Wild Game." 1, no. 4 (1924): p. 6.

Davis, H.S. "The Great Need of Fish Protection in Arkansas." 1, no. 4 (1924): p. 5.

"Department Wants to Buy Deer." 2, no. 1 (1925): p. 6.

"Father Teaches Conservation." 1, no. 4 (1924): p. 6.
Futrall, J. C. "Deer and Turkey Almost Gone, Quail Getting Scarce Says Farmer." 3, no. 1 (1926): p. 14.
"Game Brings Pleasure and Money, Says Tom Wallace." 1, no. 3 (October–November–December 1924): p. 5.
"Game Preserves Essential—Closed Season Won't Put Game Back—Preserves Will." 2, no. 4 (1925): p. 4.
Hemingway, Lance. "A Valuable Natural Resource and How It May Be Saved." 1, no. 1 (1924): pp. 13–15.
Hornaday, Dr. William T. "Wild Game and the American Pioneer." 2, no. 4, (1925): p. 14.
"How, Why and Where Have Our Fish Gone?" 3, no. 1 (1926): p. 11.
"Pennsylvania's Deer Problem." 4, no. 4 (1927): p. 15.
"Preservation and Propagation of Game." 1, no. 4 (1924): p. 14.
Rutledge, Archibald. "Raising Deer on Waste Land." 4, no. 4 (1925): p. 4.
Taylor, Gurtha. "The Advantages of Protecting and Preserving Game and Fish." 2, no. 2 (1925): p. 7.
"The Game Restoration Program." 6, no. 3 (1929): p. 12.
"The Pheasant Is Rated as a Great Game Bird." (1925).
"What Was Here Fifty Years Ago." 1, no. 1 (1924): p. 9.

Articles: *Arkansas Game and Fish*, name changed in 1992 to *Arkansas Wildlife*

"Act 124 of 1915/A Conservation Commission Is Created." 6, no. 3 (summer 1974): pp. 17–18.
Alexander, Harold. "Changing Concepts and Needs in Wildlife Management." 8, no. 4 (fall 1976): pp. 3–8.
———. "Man and Beaver—Debits and Credits." 10, no. 1 (spring 1978): pp. 12–15.
———. "The Red Wolf Saga." 14, no. 2 (fall 1983): pp. 2–6.
Armstrong, Mike. "Black Bass Length Limits: A Management Tool for Arkansas?" 17, no. 2 (March/April 1986): p. 31.
Barkley, Sam. "Eagle Eyeing in the Natural State." 18, no. 1 (January/February 1987): pp. 29–30.
———. "The American Alligator." 18, no. 3 (May/June 1987): pp. 25–26.
"Bear Encounter—What to Do." 18, no. 3 (May/June 1987): p. 11.
"Bountiful Bunnies." 1, no. 2 (winter 1967–spring 1968): p. 10.
Cartwright, Michael. "Either-Sex Deer Management." 17, no. 5 (September/October 1986): pp. 2–7.
Cartwright, Michael E. "Return of the Elk." 26, no. 4 (fall 1995): pp. 2, 4.
Cartwright, Michael, and J. Michael Pledger. "Talking Turkeys." 17, no. 2 (March/ April 1986): pp. 2–6.
Clark, Joe, and Paul Polechla. "The River Otter in Arkansas: A Success Story." 17, no. 6 (November/December 1986): pp. 6–9.
"Construction of Public Fishing Lakes." 13, no. 2 (spring/summer 1982): p. 7.
Criner, Dave. "A Helping Hand for Endangered Species." 10, no. 4 (spring 1979): pp. 23–24.

de Bin, Jerry A. "The Northern Bobwhite." 18, no. 6 (November/December 1987): pp. 27–28
Donaldson, Dave. "Honkers Avoiding Arkansas, But Why?" 2, no. 1 (fall 1968): pp. 23–26.
"[The] Eagle Reigns Supreme." 8, no. 3 (summer 1976): pp. 10–11.
Gaines, Bill. "The Eastern Wild Turkey." 5, no. 1 (spring/summer 1972): pp. 2–5.
———. "What Good Are Ducks, Deer, Bass . . ." 8, no. 1 (fall/winter 1975): pp. 18–19.
"Gators Making a Comeback." 8, no. 3 (summer 1976): pp. 4–6.
Gipson, Phil. "Coyotes and Related Canids of Arkansas." 7, no. 1 (fall 1974): pp. 8–10.
Gipson, Philip S. "The Coyote." 17, no. 1 (January/February 1986): pp. 27–28.
Gipson, Phil, and John Sealander. "Dog/Deer Relationship Probed." 6, no. 3 (summer 1974): pp. 10–13.
Gregory, Dick. "Mourning Dove." 17, no. 5 (September/October 1986): pp.27–28.
Henderson, Scott. "Don't Believe Everything You Read." 10, no. 1 (spring 1978): pp. 20–23.
Hendricks, Bryan. "Spotted Bass: Supporting Cast in Arkansas's Mountain Streams." 27, no. 1 (spring 1996): pp. 6–7.
Heuston, John. "Predator!" 20, no. 5 (September/October 1989): pp. 13–16.
Hulsey, Andrew. "Nursery Ponds Produce." 1, no. 1 (fall 1967): pp. 28–29.
Kaffka, Jay. "A New Bass for Arkansas." 10, no. 1 (spring 1978): pp. 3–5.
———. "Act 35/A New Approach to Conservation." 7, no. 1 (fall 1974): pp. 2–3.
———. "Exotics." 12, no. 3 (spring/summer 1981): pp. 2–5.
———. "Long Tales of Big Cats." 4, no. 1 (spring 1971): pp. 4–6.
———. "The Great Turkey Comeback." 10, no. 4 (spring 1979): pp. 16–17.
———. "Today's Conservation Problems Prophesized in the 1920s." 6, no. 3 (summer 1974): pp. 14–16.
———. "White River Refuge: Oasis in an Ecological Desert." 3, no. 2 (summer 1970): pp. 7–9.
Keith, Bill. "Redbreast, Cold-Water Bream." 4, no. 2 (summer 1971): pp. 10–11.
Kilburn, Marcus. "The Over-50 Club." 25, no. 3 (summer 1994): pp. 3–6.
Low, Jim. "A Survivor in Trouble." 25, no. 1 (spring 1994): pp. 26–27.
———. "Arkansas' Froggy Friends—Toadally Awesome!" 19, no. 2 (March/April 1988): pp. 11–22.
———. "Scouting Arkansas." 18, no. 5 (September/October 1987): pp. 6–7.
———. "The Armadillo." 19, no. 5 (September/October 1988): pp. 27–28.
———. "The Canada Goose." 20, no. 5 (March/April 1989): pp. 27–28.
———. "The Scales of Justice." 19, no. 6 (November/December 1988): pp. 22–23.
Mathis, Bill. "Rise and Decline of Commercial Fishing on the White River." 3, no. 1 (spring 1970): pp. 2–4.
Mosby, Joe. "Alligator Snappers Now Protected." 25, no. 1 (spring 1994): p. 27.
———. "[The] Osprey." 23, no. 4 (winter 1992): pp. 25–26.
Neal, Joe. "In the Noose of Shrinking Habitat." 12, no. 3 (spring/summer 1981): pp. 22–26.
"Our Heritage Protected." 12, no. 2 (winter 1980): pp. 5–8.
Parrish, Jim. "The Business of Hunting with Dogs." 3, no. 4 (winter 1970): pp. 15–16.
"Passenger Pigeons over Little Rock." 11, no. 2 (summer/fall 1979): pp. 12–14.
Patterson, Gregg. "Arkansas' World-Class Trout." 21, no. 1 (January/February 1990): pp. 7–11.
———. "Concern for the Bobwhite." 17, no. 1 (January/February 1986): pp. 9–13.

———. "Dark Days for Ducks . . . Is There Hope?" 19, no. 6 (November/December 1988): p. 3.
——— "[The] Great Blue Heron." 23, no. 2 (summer 1992): pp. 21–22.
———. "Research Update: Co-op Unit Is Busy." 21, no. 1 (January–February 1990): p. 20.
———. "The Largemouth Bass." 19, no. 2 (March/April 1988): pp. 27–28.
———. "Tournament Bass Fishing into the '90s." 21, no. 2 (March/April 1990): pp. 1–15.
———. "What's Happening to Our Smallmouth Streams?" 22, no. 2 (spring 1991): pp. 2–6.
Perkins, Alan. "The Snow Goose." 18, no. 1 (January/February 1987): pp. 25–26.
Pharris, Larry. "A Different Drummer for Arkansas." 14, no. 1 (spring 1983): pp. 10–12.
———. "Bear Hunting Is Tough." 12, no. 4 (fall 1981): pp. 5–7.
Pledger, Mike. "A Commissioner's Dream Come True." 8, no. 4 (fall 1976): pp. 9–11.
———. "Deer Disease Runs Its Course." 9, no. 2 (summer 1977): pp. 12–14.
———. "In Pursuit of the Declining Red Fox." 7, no. 4 (summer/fall 1975): pp. 14–16.
———. "The Beaver Is Back." 8, no. 2 (spring 1976): pp. 14–16.
———. "Varmints, Fur, and $$$." 11, no. 2 (summer/fall 1979): pp. 5–7.
"Progress Report." 1, no. 2 (winter 1967/spring 1968): pp. 14–15.
Purvis, George. "Nature's Odd One, the Armadillo." 1, no. 2 (winter 1967/spring 1968): pp. 12–13.
———. "The Impact of Amendment 35." 10, no. 4 (spring 1979): pp. 2–5.
Rogers, Mitch. "Bear Season." 11, no. 4 (summer 1980): pp. 2–4.
Rush, Gene. "Bare Facts about Bear." 1, no. 1 (fall 1967): pp. 19–21.
———. "Winona Wildlife Management Area." 5, no. 4 (spring 1973): pp. 4–6.
Saugey, David. "Aerial Acrobats of the Evening Sky." 19, no. 6 (November/December 1988): pp. 14–21.
Smith, Ronald A., and Michael E. Cartwright. "Drumming Up Success." 18, no.5 (September/October 1987): pp. 19–23.
Spencer, Jim. "At Issue: Dogs, Deer and Wise Use." 15, no. 4 (September/October 1984): p. 6.
———. "The Beaver." 19, no. 3 (May/June 1988): pp. 27–28.
———. "The Crayfish." 22, no. 2 (spring 1991): pp. 17–18.
———. "Mystery Bird." 21, no. 2 (March/April 1990): pp. 26–27.
———. "Turkey Forecast: Mixed Reviews." 21, no. 2 (March/April 1990): p. 25.
Sutton, Keith. "Arkansas' Fragile Menagerie." 17, no. 1 (January/February 1986): pp. 14–19.
———. "Arkansas Furbearers." 17, no. 6 (November/December 1986): p. 15.
———. "Arkansas Owls, Creatures of the Dark Side." 17, no. 5 (September/October 1986): pp. 19–23.
———. "Arkansas' Venomous Snakes: The Serious Six." 18, no. 3 (May/June 1987): pp. 13–20.
———. "Bobwhites and Rabbits for Tomorrow." 19, no. 3 (May/June 1988): pp. 15–18.
———. "Enchantment on the Wing: The Hawks and Falcons of Arkansas." 19, no. 1 (January/February 1988): pp. 13–20.
———. "How 'bout Them Hogs." 19, no. 1 (January/February 1988): pp. 29–30.
———. "Mud Cats, Whisker Faces and Chuckleheads." 17, no. 5 (September/October 1986, rep.
———. "The Paddlefish: Rudolph of the River." 18, no. 2 (March/April 1987): pp. 5–7.
———. "Questions Answered about New Turtle Laws." 19, no. 2 (March/April 1988): pp. 25–26.

———. "The Romance of Quail: Hunting Arkansas Bobwhite." 20, no. 6 (November/December 1989): pp. 2–7.
———. "What You Can Do for Ducks." 22, no. 1 (winter 1990–91): pp. 3–7.
———. "Will-O'-the-Wisp: The Search for Arkansas' Elusive Lions." 21, no. 1 (January/February 1990): pp. 2–6.
———. "Year-round Tactics for Stripers, Whites and Hybrids." 17, no. 2 (March/April 1986): pp.13–16.
Suzuki, Dr. Howard K. "The Role of Predators." 1, no. 3 (summer 1968): pp. 34–35.
Townsend, Lisa. "The Eastern Gray and Fox Squirrels." 17, no. 6 (November/December 1986): pp. 27–28.
"Trapping and Wildlife Management." 9, no. 3 (fall 1977): pp. 4–7.
"[The] Truth about Squirrel Hunters." 1, no. 2 (winter 1967/spring 1968): pp. 8–9.
Walter, Jim. "Reshaping an Old Tradition." 13, no. 4 (winter 1982): pp. 2–4.
"What Was Here 50 Years Ago." 6, no. 3 (summer 1974): p. 2.
"[The] Why and How of Waterfowl Reglations." 10, no. 3 (winter 1978): pp. 14–16.
Williams, Harry. "Market Hunters of Early Arkansas." 2, no. 1 (fall 1968): pp. 8–11.
Wilson, Steve N. "Help for Ducks." 19, no. 6 (November/December 1988): p. 1.
———. "Learning to Live with Leasing." 18, no. 5 (September/October 1987): p. 1.
Wilson, Steve, and Phil Gipson. "Status of Black Bear in the Western Ozarks." 8, no. 1 (fall/winter 1975): pp. 10–11.
Yaich, Scott. "Lead Poisoning: The Road to Recovery." 17, no. 1 (January/February 1986): p. 6.

Articles: *Arkansas Outdoors*

"Arkansans Learning to Live with Mercury in Fish." January 24, 1996, p. 2.
"Arkansas Shadow Bass Certified as World Record." March 20, 1996, p. 3.
"Arkansas Smallmouth Bass Management Plan Eight Years in the Making." May 17, 1995, p. 1.
"Arkansas to Be Leader in Working for Troubled Summer Songbirds." June 23, 1993, p. 2.
"Bald Eagle Nesting Activity Increases in Arkansas." June 9, 1993, p. 2.
Brader, Don. "Arkansas Hatcheries Produce over 10 Million Fish Each Year." May 24, 1995, p. 1.
"Comments on Possible Elk, White River Bear Hunts Sought." February 28, 1996, p. 2.
"Commission Increases Non-Resident Hunting License Costs." April 10, 1996, p. 3.
"Duck Outlook, Mallards Especially, Brightest in Years." June 28, 1995, p. 1.
"Ducks Plentiful, So Is Water, As Second Hunting Segment Nears." December 15, 1993, p. 1.
"Five Peregrine Falcons Prospering, One Convalescing." August 10, 1994, p. 3.
"Get Back to the Basics: Go Bream Fishing." April 24, 1996, p. 1.
"Great Blue Heron Benefits from a Mayflower Man's Chilly Swim." February 9, 1994, p. 2.
"Greers Ferry Walleye Caught in 1982 Becomes World Record." September 11, 1996, p. 1.
"Horseshoe Lake Grass Carp Surpasses State, World Records." May 10, 1995, p. 3.
"Hunters Take 124 Bears during 1995 Season." April 17, 1996, p. 1.
"Hunters Take Record Number of Deer in 1995–96." April 3, 1996, p. 1.
"Larry Red's Work at Centerton Hatchery Covered Four Decades." November 2, 1994, p. 2.

"Late Reports Push Bear Hunt to a New Record." June 19, 1996, p. 1.
"Limited Hunting of Canada Geese in Northwest Arkansas Proposed." July 2, 1992, pp. 2–3.
"Master Angler Program Provides Challenge to Fishermen." March 13, 1996, p. 2.
"More Hunting of Snow Geese Encouraged in Arkansas." July 31, 1996, p. 2.
Mosby, Joe. "Elk Viewing Is Becoming a Tourist Attraction on Buffalo River," September 30, 1992, p. 1.
Mosby, Joe. "Statistics from 1992 Show Erling 'Best' Bass Tournament Lake." May 5, 1992, p. 1.
"Nest Survey Shows Canada Geese Numbers Continue to Increase." June 4, 1992, p. 2.
Patterson, Gregg. "Cutthroat Trout Quickly Wins Popularity in Arkansas." December 9, 1992, p. 1.
"Red Fox Again Available to Arkansas Hunters, Trappers." August 7, 1996, p. 1.
"Robinwood Tract Added to Wattensaw Wildlife Management Area." March 8, 1995, p. 1.
"Saugeye: A New Fish for Arkansas Waters." October 28, 1992, p. 2.
"State Fishing Record Established for Shadow Bass." January 18, 1995, p. 2.
"Survey Shows Another High Level for Duck Populations." July 24, 1996, p. 1.
Sutton, Keith. "Crafty, Adaptable Crows Offer Challenges to Arkansas Hunters," February 16, 1994, p. 1.
———. "The Romance of Quail: Hunting Bobwhites in Arkansas." December 23, 1992, p. 1.
"Trying to Sell Snakes, Turtles Can Lead to Major Trouble." June 9, 1993, p. 2.
"Waterfowl Management Plan Celebrates 10 Years of Success." May 8, 1966, p. 2.
"Wildlife Watchers Asked to Help with Trumpeter Swan Reports." December 6, 1995, p. 1.
"Youngsters to Get Preference for Antlerless Deer Permits." April 28, 1993, p. 1.

Articles: *Arkansas Fish and Wildlife*, formerly *Arkansas Sports Scene*

Bridwell, Charles L. "Arkansas Waters Brim with Bream." May 1995, pp. 42–45.
———. "Bringing Down Doves." September 1992, pp. 22–24.
———. "Hunting the Big Arkansas Catfish." July 1993, pp. 20–22.
———. "The Life and Times of Lake Pine Bluff." June 1994, pp. 16–18, 20.
———. "White Bass Don't Jump (But They Can Really Pull)." February 1994, pp. 10–13.
Griffee, Carol. "A Rumor Worthy of the Kremlin." November 1991, pp. 10–13.
———. "Blue Mountain May Be Favoritism But It Pays Dividends to State." October 1990, pp. 17–18.
———. "Call to Political Arms for Arkansas Anglers." October 1993, pp. 36, 38–39.
——— "Gatehaters Go to Court." February 1992, pp. 13–15.
———. "Greens [*sic*] County Wildlife Club Shows the Way Again." June 1991, pp. 25–26.
———. "IP, G&FC Win Hunting Association Lawsuit." February 1993, pp. 33–35.
———. "Legislative Impact: G&FC Comes Out on Top." May 1993, pp. 36–38.
———. "Starting Over." December 1994, pp. 26–28.
———. "Swallowing Hard and Taking the Risk." August 1991, pp. 16–18, 25.
———. "Trout Fishery on the White River: The 'Symptom' Has Been Treated." March 1994, pp. 32–34.

———. "What the 'Gate Haters' Really Despise Is Change." March 1991, pp. 13–14.
Mosby, Joe. "Specklebelly Special." December 1995, pp. 20–22.
Patterson, Gregg. "Try Mountain Cats for a Change." July 1994, pp. 28–30.

Articles: *Arkansas Out-of-Doors*, all articles cited written by Carol Griffee unless otherwise credited

"Beavers Bug Lawmakers." February 1992, p. 9.
"Circuit Court Decides Game and Fish Lawsuit." January 1988, p. 17.
"Clash of Rights Goes to Court." January 1992, p. 6.
"Compromise Adopted on Importation of Coyote, Fox." May 1996, p. 1.
"Cost of Deer Hunting Put in Judge's Hands." July 1987, p. 17.
"Future of Hunting, Fishing, Trapping Is Troublesome." October 1991, p. 1.
"G&FC in 'Hold the Line' Mode as Finances Tighten." December 1989, p. 4.
"G&FC Is in Its Darkest Hour." April 1991, p. 1.
"G&FC, Soil and Water May Make '87 General Assembly Lively." January 1987, p. 8.
"G&FC to Seek Excise Tax." January 1991, p. 1.
"G&FC Urged to Be Cautious about New Lake Proposals." October 1990, p. 4.
"G&FC Wins License Fee Increase: New Trespass Law Is on the Books." March 1989, p. 19.
"G&FC Wins Race to File Lawsuit in Dispute with General Assembly." June 1987, p. 12.
"Gravel Mining Ante Raised." June 1994, p. 1.
"Gravel Mining Said to Hurt State Streams." April 1994, p. 1.
"Hostility between Legislature, G&FC Marks Session." April 1987, p. 13.
"Hunting Fee Issue Is a Political Hostage." August 1987, p. 13.
Kilburn, Marcus. "'Missouri Question' May Be Answered by Turkey Study." March 1992, p. 6.
"Land Swap Will Protect More Arkansas Wetlands." February 1993, p. 4.
"Legislators Tackle G&FC Funding Woes." March 1991, p. 1.
"Licenses, Boat Fees, Real Estate Transfer Tax Bills Creep through Legislature." March 1987, p. 4.
"License Fund Accord Is Reached (Finally)." March 1989, p. 16.
Mosby, Joe. "We're Missing Out on Something." January 1992, p. 1.
"State High Court Reverses Good Politics, Lousy Law." November 1988, p. 11.
"Stream Gravel Mining Ban Saved." May 1995, p. 1.
"Stream Gravel Miners Face New Rules by Corps, EPA." September 1994, p. 1.
"Stream Gravel Mining Fight Moves to Legislative Arena." Januuary 1995, p. 1.
"Stream Gravel Mining War Still Raging." April 1995, p. 1.
"'Take and Pay.'" March 1992, p. 1.
"Under New Management, Wetlands Program Soars." April 1996, p. 1.
"Watch Your Language!" November 1993, p. 1.
"Wetlands No More." November 1991, p. 1.
"Wetlands: Situations 'All Fouled Up.'" April 1990, pp. 4–5.

Articles: Miscellaneous Magazines

Apple, W. M. "Bill." "Alligator Gar." unknown publication. (copy contained in scrapbook kept by Apple and made available by W. M. "Bill" Apple Jr.)

Babbitt, Bruce. "Stewards of Creation." *Christian Century,* May 8, 1996, pp. 500–501.

Begley, Sharon. "Killed by Kindness." *Newsweek,* April 12, 1993, pp. 50–54.

"Finally, the Fiscal Year's Budget. *Science News,* May 11, 1996, p. 297.

Kerasote, Ted. "This Land Is Whose Land?" *Sports Afield,* September 1995, pp. 17–20.

Nowak, Dr. Ronald M. "The Mysterious Wolf of the South." *Natural History* 18, no. 1 (1972): pp. 50–51, 74–77.

Nowak, Ron. "Red Wolf: Our Most Endangered Mammal." *National Parks and Conservation Magazine,* August 1974, pp. 9–12.

van Dresser, Cleveland. "Rape of the Arkansas Bottomlands." *Sports Afield,* April 1955, p. 114.

Venters, Vic. "Return of the Red Wolf." *Wildlife in North Carolina,* September 1989, pp. 19–23.

Watts, David L. "The Wolf At Our Door." *Mississippi Outdoors,* March/April 1989, pp. 5–10.

Young, Stanley P. "It's Red But Truly All-American." *American Forests,* January 1968, pp. 14–15, 56–57.

Zbar, Jeffery D. "Environmental Marketing." *Advertising Age,* June 28, 1993, p. S–1.

Articles: *Arkansas Democrat*, Little Rock, Arkansas

"Commercial Sale of Fish Proposed as Legal." January 24, 1956.

"Fish Farms Test Suit Endorsed." February 26, 1956.

Meins, Jan. "Prairie Farmer Remembers His Pioneer Homestead." December 17, 1980.

"Wildlife Group Launch Anti-Fish Farming Move." April 28, 1957.

Article: *Arkansas Democrat-Gazette*, Little Rock, Arkansas

Lemons, Terry. "Arkansas Gains Forest after Alaska Deal Struck." October 4, 1996.

Articles: *Arkansas Gazette*, Little Rock, Arkansas

"All Members Refuse to Quit Game and Fish Commission Posts." June 1, 1967.

"Anti-Trapping Proposal Held Unconstitutional." February 12, 1956.

"Club Warden Shot the Twelfth Time." March 1, 1915.

"Cox Defends Commission Game and Fish Permits against Attacks." April 3, 1957.

Dean, Jerry. "Arkansas's Eyes of the Night—Bobcats—Are Faring Well." November 18, 1985.

"Duck Hunting In a Big Way." November 26, 1935.

"Ears Strain for Cry of Whippoorwill, Inexplicably Absent from Area." August 31, 1984.

"Egret Return to Van Buren Not Eagerly Awaited Event." Associated Press. February 14, 1986.

"Eight Acts, Amendments Filed in Time." July 7, 1944.
Fleming, John. "Duck Hunting: From Model T to Big Motels." November 26, 1972.
"Game and Fish Commission Chooses Hulsey as Assistant Director." March 26, 1968.
"Game and Fish Commission Ouster Is Appealed by Farris." November 29, 1969.
"Governor Lists Nine Charges against Farris." July 18, 1968.
"Governor Pledges to Keep Up Fight against Game and Fish Commission." October 27, 1968.
Griffee, Carol. "Ban on Dogs in Deer Hunt Spurs Outcry; Defended as Needed to Protect Wildlife." April 1, 1984.
———. "Eagle Killed, 1 Wounded; GFC Seeks Information to Help Find Attackers." January 13, 1984.
"Hackler Submits Resignation as GFC Director." May 6, 1969.
Harris, Sam G. "House Once More Beats Game Commission Fund, Other Agencies Imperiled." February 25, 1963.
Hays, Steele. "Commissioner Tells Why Hulsey Fired as G&FC Head." August 30, 1969.
"Hearings by WR Receive Clearance in GFC Ouster." May 28, 1968.
"House Flays Hackler, Twice Fails to Pass Funds for the Game and Fish Commission." March 28, 1969.
James, Lamar. "Experts Say Alligators Aren't Looking for Trouble." June 2, 1985.
"Mauling of Hunter Reported." December 21, 1985.
Mosby, Joe. "An Arkansas Farm of 'Big Birds.'" October 28, 1984.
———. "Bald Eagle Familiar Sight Now." March 2, 1986.
——— "Co-operative Work Gives Bluebirds a Helping Hand." March 3, 1985.
———. "Hulsey Fired by Commission." August 29, 1979.
———. "We Call Him Crazy, but the Ridiculed Coot Has a Certain Place." February 2, 1986.
"New Statewide Law Governs Sportsmen." March 13, 1915.
"Pollution Suit Damages Due, Ruling Holds." May 17, 1966.
"Resign, Letter from Governor Asks Game and Fish Commission." May 30, 1967.
"Reward Rises to $4,000 for Elk Killer." December 8, 1983.
Smith, Doug. "Four-Year Terms Are Approved, Casinos Are Rejected by Voters." December 23, 1984.
"Wildlife Board Clarifies Action: Cox Sole Head." May 9, 1956.
Wilson, Ruth. "Elk Returning to Native Habitat." February 11, 1985.
"WR Calls Game and Fish Commission 'Private Club.'" October 26, 1966.

Articles: *Stuttgart Standard*, Stuttgart, Arkansas

"Duck Flight Near Record." December 16, 1971. (Reprinted from unspecified Little Rock newspaper.)
"Joe Pulitzer Leases Freudenberg Reservoir." December 16, 1971. (Reprinted from unspecified publication.)
"New Precedent for Duck Hunters." December 16, 1971. (Reprinted from unspecified publication.)

Articles: Miscellaneous Newspapers

Farney, Dennis, and Timothy Noah. "Environmental Stands Alienate Some Backers of the GOP's Agenda." *Wall Street Journal,* March 5, 1996.

Greer, William R. "Near Extinction, Wild Turkeys Rebound and Head for the City." *New York Times*, November 3, 1985.

Editorials: *Arkansas Gazette*, Little Rock, Arkansas

Albright, Gus. "Hulsey's Axing and Amendment 35." September 14, 1979.

"Amendments and Acts." November 5, 1944.

"Game and Fish Regulations Should Be Flexible." October 31, 1944.

"Only One Boss." May 13, 1956.

Editorial: *Southern Standard,* Arkadelphia, Arkansas

"The Recent Game Law." Arkadelphia, Ark.: *Southern Standard*. 1885. (Provided by Dr. Larry Don Frost of Bismarck, Ark., professor of English, Henderson State University, Arkadelphia.)

Columns: *Arkansas Democrat-Gazette*, Little Rock, Arkansas

Bowman, Steve. "Arkansas River Doesn't Give Any Angler Respect." June 2, 1996.

———. "The Yanks Are Coming . . . for Walleye Pro Tournament." May 2, 1996.

Column: *Arkansas Gazette*, Little Rock, Arkansas

Albright, Gus. "Game and Fish Director Has Much to Live Up To." May 13, 1956.

Columns: *St. Louis Post-Dispatch*, St. Louis, Missouri

Coghland, Ralph. "Duck Hunting in Stuttgart." Reprinted in *The Stuttgart Standard*. Stuttgart, Ark., December 30, 1949.

———. "Trials of a Duck Hunter." December 21, 1947, Reprinted in *The Stuttgart Standard*, Stuttgart, Ark., December 9, 1971.

Columns: "Arkansas Wildlife" by W. M. "Bill" Apple, *Arkansas Democrat*, Little Rock, Arkansas

Columns Without Headlines:

December 5, 1954; March 27, 1955; July 8, 1956; July 26, 1956; September 23, 1956; December 16, 1956; June 2, 1957; June 14, 1957; December 8, 1957; August 3, 1958; September 28, 1958; October 28, 1958; April 26, 1959; May 3, 1959

Columns With Headlines:

"Bootlegger at Work." March 24, 1957.

"Deer Hunting without Dogs." September 5, 1954.

"Ducks Aren't Unlimited." October 27, 1957.

"Duck Hunters Party." October 23, 1955.

"Duck Season Promising." November 11, 1956.

"Fishing Institute Reports on Arkansas." March 6, 1955.

"Future Looks Dark for Duck Hunters." April 10, 1955.

"House Bill Would Boost License Fees." February 27, 1955.

"Hunter's Hopes High." August 14, 1955.

"In Quest of 'Big Fish'?" April 17, 1955.

"New Law Sought to Halt Shockers." July 29, 1956.

"Prospects Bright for Deer Hunters." November 13, 1955.

"Quail Hunters Prepare Plans." January 24, 1954.

"Quail Picture Looks Brighter." October 24, 1954.

"Quail Prospects Good." February 10, 1957.

"Rabbit Hunting Fun with Beagles." September 9, 1956.

"Sale of Game Fish." February 17, 1957.

"Top Fishing in Prospect." February 9, 1958.

"Trout, Turkey Seasons Open Soon." March 25, 1956.

"Walleyed Pike Neglected in Arkansas." September 4, 1955.

"Water Pollution Endangers Wildlife." April 18, 1954.

"Wildlife in Review." December 30, 1956.

Columns: Magazines

Reiger, George. "If It Ain't Broke. . . ." *Field and Stream,* June 1993, pp. 10, 12.

———. "When BIGGER'S Not BETTER." *Field and Stream,* November 1995, pp. 16–18.

———. "Who Owns Our Wildlife?" *Field and Stream,* November 1994, pp. 13–15.

Column: Miscellaneous Newspapers

Reynolds, Harry. "Reclaiming the Wild Goose Will Be Long, Tedious Job." *Memphis Commercial-Appeal.* April 18, 1958.

Interviews

Burkett, Mrs. Bennie. Conversation with author, Stuttgart, Ark., December 9, 1985.

Dean, James, ed. *Wildlife in North Carolina.* Telephone conversation with author, September 11, 1996.

Donoho, Riley N. Conversation with author, Fort Smith, Ark., December 16, 1985.

Holder, Trusten. Conversation with author, Little Rock, Ark., March 19, 1992.

Lynch, Rocky. Conversation with author, Conway, Ark., March 17, 1992.

Purvis, George M. Conversation with author, Little Rock, Ark. March 16, 1992.

Simpson, Dee. Conversation with author, Stuttgart, Ark., December 9, 1985.

Taylor, Armil. Conversation with author, Clarksville, Ark., March 17, 1992.

Miscellaneous

Act 939 of 1987. Little Rock, Ark.: Secretary of State's Office, State Capitol.

Arkansas Game and Fish Commission Minute Orders 87–08, 90–029, 90–120, 91–162, 91–186, 91–214: March 26, 1990.

Arkansas Game and Fish Commission Resolution, February 13, 1991.

Charlie Chaffin, et. al., v. Arkansas Game and Fish Commission. Arkansas Reports, 15 (1988): pp. 431–447.

"Fact Sheet." Little Rock, Arkansas: Arkansas Game and Fish Commission. undated, but probably 1984.

Hampton v Arkansas Game and Fish Commission. Arkansas Reports, May 21, 1951.

"[The] North American Wetlands Conservation Act." Fact Sheet. Arlington, Va.: North American Wetlands Conservation Council, 1996.

"Records of the States: Arkansas; Missouri Territory, 1818; Arkansas Territory, 1819–1935," Little Rock, Ark.: Arkansas History Commission. Microfilm Roll No. 1.

"Red Wolf Pups Born at St. Vincent National Wildlife Refuge." News Release. Atlanta, Ga.: U.S. Fish and Wildlife Service, May 7, 1990.

Wilson, Steve N. Statement to the Game and Fish Funding Study Committee of the Arkansas General Assembly, State Capitol, March 24, 1992.

Selected Additional Readings

Abbott, John S. C. "The Wilds of Arkansas." *Harper's New Monthly Magazine,* October 1866.

Atkinson, J. H. "Frederick Gerstaecker in Arkansas." *Pulaski County Historical Review* 27 (June 1969).

Aunard, Gustave. *The Trappers of Arkansas.* New York: T. R. Dawlerry, 1858.

Caruso, John A. *The Mississippi Valley Frontier: The Age of French Exploration and Settlement.* Indianapolis: Bobbs-Merrill, 1966.

Darby, William. *The Emigrant's Guide to the Western and Southwestern States and the Territories.* New York: Kirk and Mercein, 1818.

Dunbar, William. *The Exploration of the Red, Black, and Washita Rivers,* Boston: Houghton Mifflin, 1904.

Fleming, John. *The White River of the Ozarks; from Sabertooth Tiger to Fighting Rainbows.* New York: Gallenuile Society Publishing, 1973.

Gladson, W. N. "'Big Game' in 'American Alps.'" *Flashback,* May 1967.

Hartley, C. B. *Hunting Sports in the West, Comprising Adventures of the Most Celebrated Hunters and Trappers.* Philadelphia: J. W. Bradley, 1859.

Kelley, George P. "Hunting as a Youth in Desha County." *Desha County Historical Society,* 2 (1975).

Murphy, Edmund R. *Henry de Tonty, Fur Trader of the Mississippi.* Baltimore: The Johns Hopkins Press, 1941.

Nash, C. E. *Southern Stories.* No publisher, date given, but probably 1900. Little Rock, Ark.: Arkansas History Commission.

National Geographic Society. *Wilderness USA.* Washington, D.C.: National Geographic Society, 1973.

Paine, Albert B. *The Arkansas Bear, A Tale of Fanciful Adventures Told in Song and Story by Albert Bigelow Paine, in Pictures by Frank Ver Beck.* New York: Harper, 1925.

Porter, William R., ed. *The Big Bear of Arkansas and Other Sketches Illustrative of Characters and Incidents in the South and Southwest.* Philadelphia: Carey and Hart, 1846.

Sunder, John E. *Joshua Pilcher, Fur Trader and Indian Agent.* Norman: University of Oklahoma Press, 1968.

"Tripping Through North Arkansas in the Year 1894." *Benton County Pioneer* 4 (July 1959).

Tunnard, Christopher, and Boris Pushkarev. *Man-Made America—Chaos or Control?* New York: Harmony Books, 1979.

Turnbo, S. C. Collection of hunting stories published in newspapers in late nineteenth century. Little Rock, Arkansas: Arkansas History Commission.

Waterman, Charles E. *Hunting in America.* New York: Holt, Rinehart and Winston, 1973.

Index

A

C

D

E

F

G

M

N

O

P

Q

R

S

T

U

V

W

Y